T0297233

Trust-Based Collective View Prediction

Tiejian Luo · Su Chen
Guandong Xu · Jia Zhou

Trust-Based Collective View Prediction

Tiejian Luo
Jia Zhou
University of Chinese Academy of Sciences
Beijing
People's Republic of China

Guandong Xu
University of Technology
Sydney, NSW
Australia

Su Chen
China Mobile Research Institute
Beijing
People's Republic of China

ISBN 978-1-4899-9200-0 ISBN 978-1-4614-7202-5 (eBook)
DOI 10.1007/978-1-4614-7202-5
Springer New York Heidelberg Dordrecht London

Printed on acid-free paper

Springer is part of Springer Science+Business Media (www.springer.com)

To wife Xia Chen and daughter Chenxi Luo
From Tiejian

To wife Ms. Jane Zhu, father Mr. Sanqin Chen, and mother Ms. Xingqiao Luo
From Su

To wife Feixue Zhang and son Jack Xu
From Guandong

Preface

The task of collective view prediction is to reason about the opinion of an individual to an item by calculating his or her relevant online community's attitudes. More than 20 years' explorations in this field have made progress in developing precise and robust models and related algorithms for predicting collective view in online community. The researchers also learned the current methods' advantage and understood their limitations. Our research presents a new perspective and ideas to address the low performance and robustness in collective view prediction tasks. The applications of the models, methods, and algorithms in this book will be promising and valuable for improving the quality of online information recommendation services, targeting advertisement delivery, word-of-mouth analysis and so on.

Investigating related theory and engineering practice can help us understand the pros and cons of conventional methods in completing the group opinion prediction task. Recommendation methods and sentiment analysis are closely related to the task. Collaborative filtering is such a typical approach. The advantage of this method can generate personalized predictions without additional text analysis. Those approaches have become mainstream models and methods for collective view prediction in recent years. Early study indicated that the feasibility of collaborative filtering is solely based on the reliability of information resources. However, in the real application environment, things are getting complicated. The misconduct behaviors in rating could decrease the information reliability. Those activities could make the prediction invalid and unreliable. To improve the prediction accuracy and reduce the impact of noise data, the theme of this book is to review the previous related theoretical foundation and propose a trust-based collective view prediction model and relevant algorithms. Our study shows that effective model for collective view prediction is attributed to users' trust relationships network.

Asking appropriate research questions could motivate us to pursue the right direction and address hard problems in the right way. From the theoretical perspective, this book re-examines the trust definition and quantitatively analyze the relationship between user's similarity and their trust network leads us to the right solutions. From algorithm design perspective, one of the key questions is what kinds of trust metrics strategies would impact the collective view prediction accuracy. From the evaluation perspective, we establish a framework for assessing the model's robustness and to formally describe the attacks aimed at trust-based prediction algorithms.

This book studies on the linear correlation of trust and similarity, and the influence of spread distance to the correlation. To explore the trust network, we collect more than 300,000 users' data from the popular review websites. The study results indicate that users' similarity on opinions is positively correlated to their distance in trust network and negatively correlated to their trust value. We conclude two basic rules that are important to designing effective and efficient collective view prediction algorithms. In order to analyze how different trust metrics influence the prediction accuracy, we further elaborate on two well-known trust metrics, and based on the new metrics we design new collective view prediction algorithms. To further improve the accuracy of the trust-based prediction algorithms, we propose a Bayesian fusion model for combining trust and similarity. Moreover, a second-order Markov random walk model is proposed to alleviate the sparse data problem in similarity measurement. These new approaches are more accurate than the classical collaborative filtering algorithms in our experimental evaluations.

Trust-based collective view prediction demonstrates more capability to resist attacks over the traditional techniques. But there were few quantitative analysis of this issue in previous studies. We build a robustness analysis framework to measure the capability of trust-based prediction algorithms to resist attacks. Simulation results using this framework reveal the key factors which impact the robustness of trust-based algorithms, and confirm that the honest users' feedbacks can help algorithm recover from attacks. We also give two strategies to improve the algorithm robustness in real applications.

Keywords Collective view · Trust metrics · Trust network · Social network · Sentimental analysis · Recommendation · Collaborative filtering

Contents

Chapter 1
Introduction

1.1 Background

The rapid growth of the Internet has profoundly changed the way people communicate with each other and gather information. Nowadays, more and more people are accustomed to purchase on eBay, download data from Peer-to-Peer network, share videos on YouTube, publish and read reviews on ePinions.com, participate in social activities on Facebook and learn through Wikipedia. The key value of these types of applications is group participation and interaction. O'Reily defined them as Web 2.0 in order to differentiate with the other applications which mainly rely on the content providers, such as news sites and digital libraries. The new era of the Internet has become the extension and development of human society. Billions of people carry out transactions, participate in social activities and share resources on the Internet, constituting a cyberspace which contains rich emotions and insights.

Under this new pattern of information production-consumption, contents and reviews generated by public spread rapidly through the Web, and have a significant impact on people's social life and opinion formation. According to the investigation report supported by Convergys which is a famous customer management services company, a negative comment on the social networking websites may result in loss of up to 39% potential customers. In a survey of 2,000 UK consumers, one-third of respondents said they will publish their bad shopping experience on the web. The report also pointed out that the websites and videos are fostering a new phenomenon called "silent drain", which means the user who saw the negative comments will change their choices without any complaint. The power of social reviews has driven the academia and the industry to study how to collect these structured or unstructured data from BBS, Blogs, product review websites, and how to mining the collective view of users to a specific product. The process is called collective view[1] analysis or word-of-mouth analysis.

---

[1] In this book, we interchangeablely use the term "view" and "opinion".

T. Luo et al., *Trust-Based Collective View Prediction*,
DOI: 10.1007/978-1-4614-7202-5_1, © Springer Science+Business Media New York 2013

Collective view is important for enterprises, consumers and administrative organizations. For enterprises, it is helpful to improve product and service quality; for the general public, it helps to know the pros and cons of product and services they consider to buy better; for the government, it helps to understand the formation and direction of the public opinion, adjust public policies and respond to emergencies better. After nearly a decade of development, there have been some mature Internet companies whose main business is collective view analysis, such as Radian 6,[2] Visible Techniques,[3] Biz360[4] abroad and IMonitor,[5] BenGuo[6] in China. The Daqi.com[7] also provides a free service for product opinion analysis. From the technical perspective, the most widely used method to get numerical results from unstructured text data is sentiment analysis technology, which is based on natural language processing. Product review website, such as Epinions.com and DianPing.com in China usually define the metadata of reviewed objects, and then ask the users to rate for one or more attributes, and finally get the conclusion through weighted aggregation. This type of applications takes user ratings to indicate their points of view, which omit the text analysis procedure. We have to point that numerical form and text form of user reviews is logically equivalent for collective view analysis, although the analysis based on text data is more difficult.

Collective view analysis can get comments from one or more web communities, or even the entire Internet, which greatly expanded the way consumers and enterprises gather information. However, this change also brought some negative impact. Compared to the information obtained from friends or questionnaire survey, the reliability of data which collective view analysis depends on is doubtful. As more and more individuals and enterprises have realized the power of word-of-mouth, writing fake reviews to attract consumers has been becoming a usual marketing tool. Figure 1.1 shows the employment advertisement for fake comments writers released by a well-known computer equipment supplier Belkin. Employees will be paid 65 cents for each fake review written on Amazon.com which meet the requirements. The fake review writers should not only make their own positive comments seem "Real", but also give "Not Helpful" comments to those reviews which make negative comments on the same product to reduce their impact to the readers. Similar examples include a technical failure on Amazon in February 2004, resulting in the identity leak of thousands anonymous book reviewers. Surprisingly, John Rechy, who is the author of the bestselling novel in 1963 named "City of Night" and the winner of PEN-USA-West's Lifetime Achievement Award, wrote comments to his own book anonymously and give

[2] www.radian6.com.

[3] www.visibletechnologies.com.

[4] www.biz360.com.

[5] www.iwomoni.com.

[6] www.ibenguo.cn.

[7] www.daqi.com.

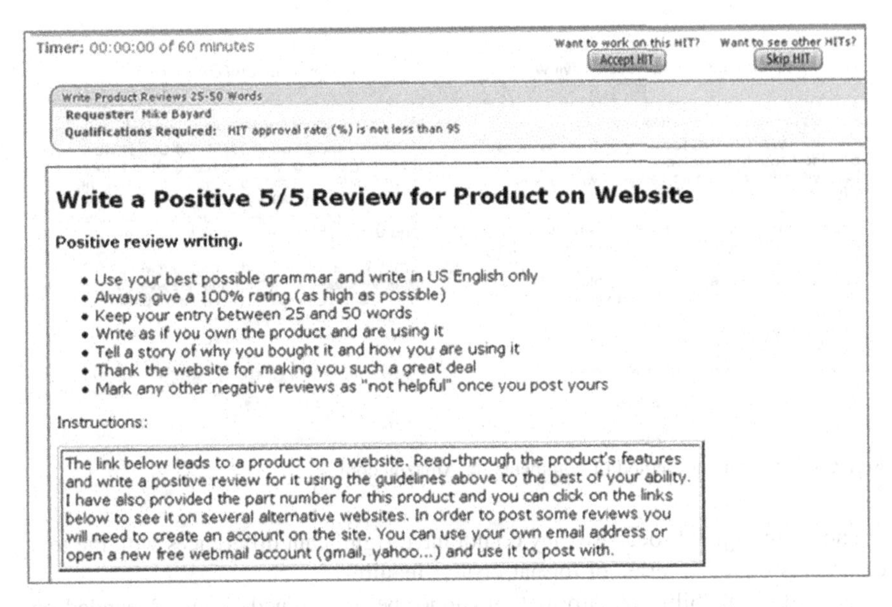

Fig. 1.1 Employment AD for fake comments releasers [1]

five-star (the highest rating level) on Amazon. The famous author Walt Whitman and Anthony Burges also have done the same thing. From these examples, we know that although the network comments have an important influence on our daily life, some individuals and enterprises also use fake reviews to make profits. Therefore, there is an urgent need to get a reliable collective view by mining and filtering the comments data.

There are some similarities between reliability evaluation of comments and spam detection, they both classify a piece of information into good or bad based on certain criteria. Rule-based filtering and text features-based filtering are two common methods for spam detection. The former determines whether a piece of information is abnormal by setting a series of rules, such as the font, usage of capital letters, specific keywords related to medicines, colors of HTML elements and so on. Rule-based filtering methods are easily identified by the attackers, and then take corresponding countermeasures. In addition, the reasonability of rules is influenced by the application environment. For example, a filtering rule which is suitable for mail client may not be appreciate to product comment website, for product comments usually contain some keywords which are the sensitive advertisement words in e-mail. Text features-based filtering is a more advanced method, which is usually based on supervised learning. It trains the classifier using the spam samples, and then determines whether a specific piece of information is spam based on the text feature. In this way, the classifier can get higher accuracy by introducing more training data, and can be applied to different application situations. However, these two methods cannot identify well-written fake reviews

Fig. 1.2 The ranking method of comments on Amazon.com

as shown in Fig. 1.1, because they do not have significant differences with the real comments either in terms of format or text features.

Since the reliability of information cannot be determined simply depended on the format and text features, there is some research work to study how to determine the reliability based on the fact contained in the information. For specific information consumer, if the fact contained in a comment is consistent with his experience, then the information is reliable for him. So we can determine the reliability of the information by collecting and mining users' feedback of reviews. However, different people may have absolutely different feedback to the same information. As shown in Fig. 1.2, Amazon sorts the comments based on their categories and the number of support. People often form two opposing groups for the same product, that is to say, each kind of point has many supporters (the comments are "helpful"). In this popularity-based filtering method, attackers can get a higher ranking of the favor comments by giving multiple votes. Similarly, this kind of multiple votes attack can also impact the results of collective view analysis. In addition, it is questionable to show the same results to all users, because different individuals have different views.

Collaborative filtering (CF) [2] provides another effective way to determine the reliability of information. It assumes that people's opinions can be represented by their historical ratings and comments. Similar users can be found by measuring the similarity of their historical ratings. Due to the stability of people's view, the opinions of the similar users are closer to the target user's thoughts. The personal analysis results for the target user can be generated by aggregating the similar users' ratings comprehensively. The input of CF is the ratings of the user group, and the output is the predicted results for a target user to his unknown items. It not only considers the ratings, but also takes the similarity of the members in the group with the target user into account. This process is to make prediction of the user's further behavior based on his known rating history. So we call this process as "Collective View/Opinion Prediction" to distinguish it from "Collective View/Opinion Analysis".

CF methods have been widely applied on many large web sites, including Amazon.com, Google News and Netflix.com etc., because their independence of content and low cost to implement [3]. Empirical evaluations have shown that this technique is effective to increase the click rate [4] and the revenue [5]. However, as pointed by many literatures, CF also has some limitations [3] and robustness [6, 7] problem. First, the users can receive satisfactory services only after providing sufficient feedbacks. This limitation is called "cold start" problem. Second, the dataset is very sparse in the practical application, so it is difficult to measure similarity effectively, resulting in low prediction accuracy and coverage. Some researchers proposed hybrid methods which combine content-based and CF technique to solve the above two shortcomings. The approaches build denser rating matrix by extracting the text information from contents [8], or measure similarity by using the "transitive relevance" [9]. Finally, the calculation of similarity in CF depends on the relevance of users' historical rating. Therefore, attackers can control the algorithm results by creating fake ratings when they knew the similarity calculation formula [6]. Although some researchers have proposed several improvements for CF to solve the robustness problem in recent years [10-13], these methods are mostly compute-intensive and require to process dataset offline, as the literature [14] pointed out. Additionally, these methods often require setting some complex model parameters, such as the threshold of noise filtering. Usually, the process of parameter optimization is complex and may have the over fitting problem.

A reliable collective view prediction method should meet the requirements of accuracy and robustness. It is not only able to mine the hidden knowledge from the group's historical behaviors, and provide personalized prediction results, but also can effectively filter noise data under attack. As discussed above, it is difficult to determine whether a piece of information is noisy only depending on the text features. Moreover, making use of users' feedbacks on reviews or CF method is easily influenced by the fake comments and ratings. Therefore, to solve this problem, we need an approach to judge the reliability of information providers, and give higher weight to credible users' ratings when predict the final results. In order to provide personal prediction results, the trust measurement should also be done personally.

To reach this goal, two basic elements are needed: trust relationship data and algorithms. Trust relationship data reflects the judgment of each other's historical behaviors in the group, such as "Jack thought that Lily's comments are helpful to him". According to our survey, there are two main ways to collect trust relationship data on the Web. Firstly, users can rate others' comments. The rating reflects the attitude of the current user to the author of the comment. We can use statistical method to build the trust relationship based on those rating information. Secondly, the current user rates other users directly. This kind of relationship is called "friend" in Dianping.com,[8] while on Epinions.com it is named as

[8] www.dianping.com, the largest review website in China.

"trust statement". A special kind of social network called "trust network or trust web" can be built by collecting such information, which reflects the mutual trust between users. The job for a prediction algorithm is: given the target user u_x, the unknown item i_y, the user collection U_y who have rated i_y, and the trust web G, the algorithm gives weight to each user in U_y according to their relationship in G, and makes the prediction by weighted aggregation.

1.2 Research Theme

The process of trust-based collective view prediction is shown in Fig. 1.3. Comparing with the CF, trust-based collective view prediction takes trust network as the input of neighbor selection and weight calculation, and then generates prediction results. In other words, trust-based collective view prediction makes better use of users' experiences and knowledge, not just mining the relations of ratings.

In recent years, the idea of trust-based prediction has been adopted in some literatures [15–17]. Studies have shown that the new approach which replaces the similarity estimation with trust metric can get higher prediction accuracy and coverage comparing with the traditional methods. However, these studies still did not answer the following basic questions:

1. What is the relationship between trust and the similarity of users' opinions? We note that the word "trust" is also used in some CF literatures [18–20], in which

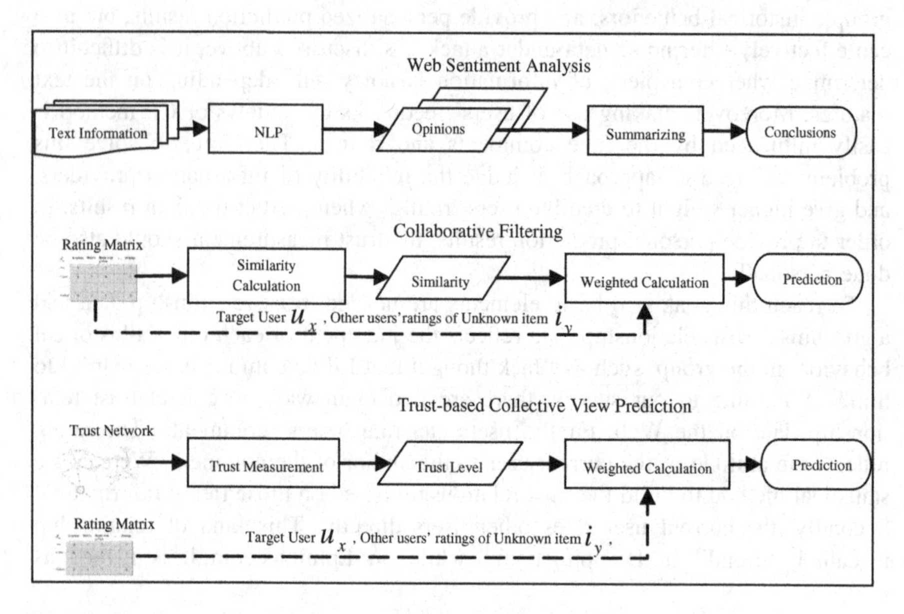

Fig. 1.3 The process of web sentiment analysis, collaborative filtering and trust-based collective view prediction

"trust" is actually the alias of user similarity or reputation. In these studies, researchers also use the transitive relevance to solve the data sparse problem. Therefore, in order to lay the sound theoretical foundation, it is necessary to clearly define the concept of trust and user similarity. In addition, there are a variety of trust metrics methods for neighbor selection and weights calculation in the prediction algorithm, but these methods are mostly heuristic. Which method is more effective for generating better prediction results? To answer the above question, the relationship between the structure of trust network and users' opinions should be well understood. The research results will become the basis of designing prediction algorithm.

2. How about the robustness of trust-based prediction method when faced with attack? Due to the highly dependencies on the topology of trust network, the attackers can control some nodes in the network through various methods in order to impact the output of prediction. How to describe the potential attacks and the ability to prevent attack of the system? What elements will affect the ability to resist attack of the system? What strategies can be used to improve the robustness? This book will present a formal assessment framework, and give answers to these questions by conducting simulation experiments based on real data.

1.3 Scope and Strategy

In Chap. 2, we will review the four research areas which are related to the book first. (1) Recommendation engines which generate personalized predictions;(2) Sentiment analysis which gets collective view from web text;(3) Studies on the regular pattern of web, including the evolution laws of social network, analysis and modeling of network information cascades;(4) Other related research in computing science which are relevant to the term "trust", mainly including federated authentication, trusted payment platform, reputation system and so on. These four areas constitute the background of this book and they are also the origin of the methods in this book.

In Chap. 3, we will describe collaborative filtering recommendation method. Algorithms for collaborative filtering can be grouped into four general classes: neighborhood-based method, latent factor model, graph-based model and socialization recommendation. We will introduce the concepts and main algorithms of them in this chapter.

In Chap. 4, we will describe sentiment analysis in detail. The basic tasks of sentiment analysis mainly includes: (1) sentiment identification, which aims to identify whether a piece of text expresses opinions; and (2) sentiment orientation classification, which aims to determine the orientation of an opinionated text. We will briefly review some methods on the above two tasks.

Table 1.1 Research phases

Phases	Questions to answer
Definition and analysis	What is the definition of trust?
	How can we get trust information on the web?
	What are the benefits it will bring to collective prediction?
	Can we quantitatively analyze the relationship between similarity and the structure of trust network?
Algorithm design	How to design collective view prediction algorithm based on the result of the above process?
	How do the different trust metrics strategies impact the prediction accuracy?
	Can we combine similarity and trust metric to improve prediction accuracy?
Evaluation	How to formally describe the attacks aimed at trust-based prediction algorithms?
	How to quantitatively evaluate the robustness of trust-based prediction algorithm?
	How to improve the robustness of trust-based prediction algorithm?

In Chap. 5, we will describe our work following the steps shown in Table 1.1. Firstly, we will define trust and user similarity clearly at concept level and explain their differences and relations. Then, we will qualitatively analysis the benefit of replacing similarity measurement with trust metrics for collective view prediction. On this basis, we will turn to experiment. The data sets we used are from Epinion.com and DianPing.com. We will present the relationship between the structure of trust network and collective view, and the differences of trust networks which are built in different ways. The analysis results in this chapter are the basis of designing the prediction algorithms.

The research methodology follows three major phases. Each phase have several specific steps which will answer some questions are relative to the main research theme. Table 1.1 lists the relevant scientific questions should be answered during the research endeavor.

In Chap. 6, we will propose two perdition algorithms based on breadth-first search algorithm, whose core ideas are weighted majority and spread of activation strategy respectively. They both belong to neighbor learning algorithm according to the classification of machine learning theory [21]. We will reveal the pros and cons on prediction accuracy of these two strategies through experiment. We also explore the prediction method combining trust and user similarity, which uses Bayesian fusion model to get higher prediction accuracy. Moreover, we propose a CF algorithm based on second-order Markov process in order to solve the sparse problem faced by traditional methods.

In Chap. 7, we study on the robustness problem of trust-based prediction algorithm. We give the formal description of attack, attack overhead, algorithm robustness and attack strategies. Then we conduct quantitative analysis of the proposed approach, finding the key elements which influence their robustness and presenting the methods to enhance their robustness in practical applications.

Finally, we present the main conclusions and the future work.

1.4 Contributions

The first research on the relationship between user similarity and trust was discussed in Dr. Ziegler's book [17]. He analyzed the influence of presence and absence of trust statement on user similarity quantitatively. This book further studied on the linear correlation of trust and similarity, and the influence of spread distance on the correlation. The analysis results show that the similarity of user opinions negatively correlated to their distance in the trust network, and positively correlated to their trust degree with each other. Then we conclude two basic rules which are important to design the effective and efficient algorithms in the collective view prediction task. They are nearest neighbor rule and similarity weighted rule. Moreover, our work is different from other former research [15–17, 22], which has taken into account only one factor, that is, users' trust statement for building trust network. Our experiments conducted based on Epinion.com dataset in this work present that using user' s mutual ratings comments as input and build trust network based on Beta distribution function can more accurately predict results than the real trust network constructed only from users' trust statement. The trust network built in this way contains more user nodes, and thus it is helpful to enhance the coverage of trust-based prediction algorithm.

In order to analyze how different trust metrics influence the prediction accuracy, we further improve two well-known trust metrics which are Mole Trust [15] and Appleseed [16], and based on the new metrics wc design new collective view prediction algorithms. The experimental results reveal that the trust-based algorithms are better than the typical collaborative filtering algorithms in terms of prediction accuracy and prediction coverage. Furthermore, experiments show that the trust metrics strategy has little effect on the predition results. Additionally, we present a similarity measurement method based on second-order Markov chain and propose a Bayesian fitting model to combine it with the trust-based prediction algorithm. Experimental result show that the new methods can achieve better prediction accuracy and coverage comparing with the traditional CF approaches.

The capability to resist attacks is a significant advantage of trust-based collective view prediction methods comparing with the traditional techniques. But there were few quantitative analyses of this issue in previous studies. In this book, we propose the formal definition of the strength of attack, metrics of robustness, and summarize the commonly used attack strategies. The experimental results show that the algorithms we proposed can recover from the attack by taking advantage of user feedback. In addition, we present the key factors which influence the robustness of trust-based prediction algorithm, and propose the strategies to improve the robustness of algorithm in practical applications.

Chapter 2
Related Work

In group participated web applications, comments and ratings made by users can be collected by system owner or web spider program. This information reflects people's opinions on specific objects. Mining this information can produce new knowledge that individual cannot provide in the past. In this chapter, we first review two research areas related to user opinion analysis and prediction, including recommendation algorithm and sentiment analysis.

The interaction of different people in social network forms several types of relationship, by which the network constructed shows some interesting statistical properties and evolution laws. The structure of the network has a significant impact on the propagation of information. These properties and laws provide theoretical basis and methodology support to make prediction on collective view through trust network. Thus, we will also review some related research on dynamic network mechanism in this chapter.

In the field of computer science, trust is a widely used term. Mainly related research sub-areas include: federated authentication, credible third party payment platform, reputation system. Jøsang et al. [23] summarized these research sub-areas as "trust management" research area. In the end of this chapter, we will give a brief review on these research sub-areas and describe the relationship between them and the research work in this book.

2.1 Recommendation Algorithm

Recommendation algorithm provides sorting prediction and rating prediction through the analysis of users' preference.

By analyzing users' preference, sorting prediction aims at providing a recommendation list with limited length for specific user from mass information to improve user's experience. Given user x and its request q, content set I, length of content list l, sorting prediction filters out the content set I_s which satisfies specific condition:

T. Luo et al., *Trust-Based Collective View Prediction*,
DOI: 10.1007/978-1-4614-7202-5_2, © Springer Science+Business Media New York 2013

$$I_s = \arg max_{I_s \in I \wedge |I_s| = l} \sum_{i \in I_s} P(i \text{ is relevant} | u_x, q) \quad (2.1)$$

and makes it most relevant to the target user. Sorting prediction can be divided into free prediction and conditional prediction according to the scope. The former is to filter among the whole content set without clear request. For example, Google provides news recommendation service. The latter is to filter among the content subset which is relevant to user's specific query. For example, in the movie rating website, the system recommends movie list according to the type of movie selected by user.

Sorting prediction can help users find their interested information, but it is difficult to evaluate the quality of the information. So in many web applications, users are allowed to rate the content. Rating is expressed by internal value and every value represents a clear semantic. For example, 1 means very bad, 5 means very good. Rating prediction is used to complete the following task: given user x and an unrated object y to determine what the possible rating is. In probability, this can be expressed as calculating the probabilities of each rating value and taking the rating with the maximum probability as the output, as Eq. (2.2) shows.

$$\rho'(u_x, i_y) = \arg Max_v P(\rho(u_x, i_y) = v | u_x) \quad (2.2)$$

As rating is always expressed by consecutive integers, it is not a good choice to treat each different value as an independent class label. Thus, expectation form like Eq. (2.3) is always adopted in literatures to do the weighted calculation.

$$\rho'(u_x, i_y) = \sum_v v \cdot P(\rho(u_x, i_y) = v | u_x) \quad . (2.3)$$

In practical Web systems, rating prediction and sorting prediction are always combined with each other. The system usually provides a recommendation list and makes rating prediction on each item in the list. For example, in product rating website, the system usually recommends product for users and provides a comprehensive rating on the product. When combing rating prediction with sorting prediction, we should consider the priority problem of them. When sorting prediction gets higher priority, it may appear that items in the recommendation list have high relevance but the predicted rating is low. In the application of product recommendation, it doesn't make any sense to recommend bad product to users. When rating prediction gets higher priority, it may appear that items in the recommendation list have low relevance. This situation should also be avoided. A possible solution is to do sorting prediction first and set a threshold value for the rating of the items in the list. Thus, only item with rating which is higher than the threshold will be displayed.

Both sorting prediction and rating prediction come down to the calculation of the utility value $\rho'(u_x, i_y)$ of a recommendation candidate to a target user [7]. In sorting prediction, utility value corresponds to the relevance degree. While in rating prediction, utility value corresponds to a specific rating value.

Recommender systems are usually classified into three categories, based on how recommendations are made: content-based methods, collaborative filtering (CF) methods, and hybrid approaches. Content-based recommendation calculates the utility value based on the target user's historical visited data and the text features of the candidate items for recommendation. Collaborative filtering (CF) tries to find desired items based on the preference of the set of similar users. In order to find out like-minded users, it compares other user's ratings with the target user's ratings. And then the target user will be recommended items that people with similar tastes and preferences liked in the past. It is not necessary to analyze the contents of items, therefore, it can be applied to many kind of domains there a textual description is not available. Hybrid approach combines collaborative and content-based methods, which helps to avoid certain limitations of content-based and collaborative systems. In this section, we will introduce content-based method and hybrid approach in brief. In the next section, we will introduce CF method in detail, due to its successful using in business and highly correlation to our work.

2.1.1 Content-Based Recommendation

Content-based recommendation calculates the utility value based on the target user's historical visit data and the text features of recommended items. These kinds of approaches can be divided into two categories, instance-based learning algorithm and model-based learning algorithm. Both of them need to express user's preference and recommended candidates as the vector of attributes.

We can get the attribute vector of recommended candidates by text processing technology or by manual labeling (For example, Web 2.0 websites widely adopt the Tagging function). TF-IDF [24] (term frequency/inverse document frequency) is a representative algorithm for text factorization. Let $\vec{\omega}_y = (\omega_{y1}, \omega_{y2}, \ldots \omega_{yl})$ represents the attribute vector of the recommended text item i_y, in which ω_{yl} represents the lth attribute's weight. Suppose that the number of texts can be recommended is N, and keyword k_l is occurred in n_l texts. Let f_{yl} represent the times of keyword k_l occurred in the text i_y, then the term frequency TF_{yl} is defined as follows:

$$TF_{yl} = \frac{f_{yl}}{\max_{l} f_{yl}} \tag{2.4}$$

The max function returns the maximum number of occurrences of all the key words in text i_y. Besides, as keywords which occurs frequently in all documents bring little information, IDF_l is used to weight TF_{yl}:

$$IDF_l = \log \frac{N}{n_l} \tag{2.5}$$

$$\omega_{yl} = TF_{yl} \times IDF_l \tag{2.6}$$

In practical applications, the text is usually represented by a certain number of keywords. For example, in Fab [25] system, 100 key words with highest weight are used to represent a web page. Given target user u_x, we can calculate all the attribute vectors of texts created or rated by the user with Eqs. (2.4–2.6). Combining with information of user rating, we can get the vector representation of user preference: $\vec{\omega}_x = (\omega_{x1}, \omega_{x2}, \ldots \omega_{xl})$. There are many kinds of approaches to calculate the $\vec{\omega}_u$, such as simple weighted calculation, Rocchio [26] and Winnow [27] algorithm. After getting the attribute vector representation of user preference, we can get the utility value of text by comparing it with the content vector of text which can be recommended. Cosine measure is a commonly used similarity measurement. The specific formula of cosine measure is as follows:

$$\rho'(u_x, i_y) = \cos(\vec{\omega}_x, \vec{\omega}_y) = \frac{\vec{\omega}_x \cdot \vec{\omega}_y}{|\vec{\omega}_x| \cdot |\vec{\omega}_y|} \tag{2.7}$$

Model-based approaches archive each user's prediction model through training sample set. Multiple attribute values of the information are taken as the input, and the utility value is taken as the output. Typical model-based algorithms include Bayesian classification [28, 29], clustering, neural network, decision tree [29] and support vector machine [30]. A novel prediction method is proposed in [31], it generalizes a related model from user click behavior and web page data to recommend web page. Although the model is content-based, it implements content decoupling by ingenious model generalization and can recommend new web pages different from the user concerned before.

Content based recommendation system mainly has the following disadvantages [3]: First, attribute extraction problem. A piece of information must be expressed as a certain number of attributes or keywords. We can automatically complete this job for text type information (news, web pages) by keywords extraction techniques. But for other types of information, such as movie, music, product, it is impossible to automatically get an ideal attribute representation and manual attribute definition. This requires a lot of domain knowledge for system designers and system maintenance worker, which bring extra knowledge acquisition burden. The second problem comes with the recommendation scope. As only items similar to the past preference of the user will be recommended, items not similar to user's preference will not be recommended even if the user may be interested in these items. To solve this problem, some recommendation systems introduce randomness. The last problem comes with new users. The recommendation system can calculate out new user's preference only after he has provided enough ratings. Thus new user will not get ideal recommendation results in the initial period, thereby affecting their willingness to use the system.

2.1.2 Collaborative Filtering

By measuring the user's access history (include download, purchase, click, rating), collaborative filtering recommend the target user with new items preferred by other users similar to the target user. This research began in the beginning of last century. Resnick et al. [32] built a large-scale collaborative filtering system in Grouplens to recommend high quality news to users. Compared to content-based recommendation, collaborative filtering doesn't require semantic analysis on the recommended content, and overcoming the limitation of recommendation scope in content-based recommendation [3]. Thus, collaborative filtering is widely used in many large-scale commercial systems, including Google News [4], Netflix.com and Amazon.com [5]. Some researchers [18, 19] use the term "trust" clearly in the description of their algorithm, whose main idea is the use of trust transitivity to solve the sparse data problem. As these algorithms only consider the correlation of users' historical ratings, the term 'similarity' may be a more appropriate description.

Algorithms for collaborative recommendation can be grouped into four general classes: neighborhood-base method, latent factor model, graph-based model and socialization recommendation. Because this part of content is highly correlated to this book, we will introduce it in detail in the next chapter.

2.1.3 Hybrid Methods

Several recommendation systems use a hybrid approach by combining collaborative and content-based methods, which helps to avoid certain limitations of content-based and collaborative systems [25, 33–38]. Different ways to combine collaborative and content-based methods into a hybrid recommender system can be classified into four types.

1. Implementing collaborative and content-based methods separately and combining their predictions.

In this approach, first we can combine the outputs or ratings obtained from individual recommendation systems into one final recommendation using either a linear combination of ratings [34] or a voting mechanism [35]. Alternatively, we can choose one of the individual recommenders which is perform "better" than others based on some recommendation "quality" [39, 40].

2. Adding content-based characteristics to collaborative models.

Several hybrid recommendation systems are based on collaborative methods but also maintain the content-based profiles for each user. These content-based profiles are used to calculate the similarity between two users. This kind of technique can overcome some sparsity problems of a purely collaborative method since not many pairs of users had a significant number of commonly rated items [35]. Reference [41] employs an approach in using the variety of different content-analysis agents that acts as additional participant in a collaborative filtering community. As a result, the users whose ratings agree with some of the agents' ratings would be able to receive better recommendations.

3. Incorporating some collaborative characteristics into a content-based approach.

   The most common and popular approach in this category is to user some dimensionality reduction techniques on content-based profiles. Reference [42] uses latent semantic indexing to create a collaborative view of a collection of user profiles, this results in a performance improvement compared to the pure content-based method.

4. Constructing a general unifying model that incorporates both content-based and collaborative characteristics.

   This kind of methods is very popular in recent years. Reference [33] proposed using content-based collaborative characteristics in a single rule-based classifier. References [43] and [44] proposed a unified probabilistic method for combining collaborative and content-based recommendations, which is based on probabilistic latent semantic analysis. Reference [45] uses Bayesian mixed-effects regression models to employ Markov chain Monte Carlo methods for parameter estimation and prediction.

2.2 Sentiment Analysis

Textual data contains two main kinds of information: facts and the emotions. Research area such as search engine and text mining focuses on exploring facts from the text, while sentiment analysis studies on the computation methods of opinions, moods and emotions from the text [46].

The rapid development of web applications makes it easy to obtain amounts of text information which contains the users' opinions. All these information are distributed in the application systems such as BBS, social networks, blogs and review sites and so on. Collecting this information by crawling and doing sentiment analysis, we can get the objects' (e.g., products) web word-of-mouth. The manufacturers can know about the users' basic impression on their products, the products' features which are satisfactory or dissatisfactory, then they can improve the products or deal with crisis management based on these knowledge.

Due to the difference of granularity, sentiment analysis is classified as document level, sentence level and feature level [47]. Document level classified one comment as ve+ and ve−. The user's comments can be translated to a corresponding value by document level analysis for further analysis. Sentence level classifies any single sentence in the document as objective fact and subjective attitude and maps every subjective attitude to ve+ or ve−. Although document level or sentence level analysis has its own emphasis, we don't have enough commentary to find out what the user like or dislike from one comment. Feature level can help us obtain the main opinion which the user has for the main characteristic of reviewed object.

The main technique used in sentiment analysis is natural language understanding [1]. How to map human's abundant emotional expression to numerical value accurately is still a challenging task. We have to point out that sentiment analysis will also be affected by spam [48]. As described in Chap. 1, identifying spam simply depending on the text features is hard. At present, there isn't any perfect solution for this problem [46]. Finally, sentiment analysis focuses on obtaining structured expression of user's attitude from un-structured comments. Certain tasks, such as recommending some contents which the user may be most interested in based on his historical comments or ratings, or generating personalized group opinion analysis results based on the subjective preferences of the user, are beyond the main scope of sentiment analysis. They all belong to the topics of recommendation engines mentioned above. Because this part of content is highly correlated to this book, we will introduce it in detail in the Chap. 4.

2.3 Dynamic Network Mechanism

Users' interaction in the web systems forms a network which contains different types of relationships. These networks show significantly different statistical properties and evolution rules comparing with a randomly generated network. Grasp and application of these rules will directly affects the design and evaluation of collective view prediction algorithm which takes trust network as input.

Strictly speaking, the network N is a connected, acyclic, directed graph which meets the following requirements: (1) there is a subset of vertices X and the in-degree of each node is 0; (2) there is a subset of vertices Y which is disjoint with X, and the out-degree of each node in Y is 0; (3) each edge has a non-negative weights, called edge capacity. The graph G is a two-tuple (V, E), where V is the set of nodes, and E is a subset of $V \times V$ called edge set. If the element in E is ordered, then G is a directed graph. If the element in E is unordered, namely $(i,j) \in E \Leftrightarrow (j,i) \in E$, then G is an undirected graph. As people often use the term 'network' as the synonymous of 'graph' in research area, in this book we take 'network' and 'graph' as the same term.

2.3.1 *Statistic Characteristics*

There have been extensive studies on the statistical properties of various types of network, including social network, cited network, food chain network, Internet and so on. The results show that, although there are many differences in scale and filed, most networks follow some common rules. The most important three properties are "long tail distribution", "small diameter" and "clustering effect".

2.3.1.1 Long Tail Distribution

Long tail distribution means the number of nodes whose degree (out-degree and in-degree in directed graph) is d, denoted as N_d, follows the power law distribution $N_d \propto d^{-\gamma}$, and γ is the power rate index [49]. This kind of distribution has been observed in phone call network [50], web pages link network [51–55], the Internet [56], cited network [57], web social networks [58], and other research filed. The study also found that the parameter γ is generally between 2 and 3. For example, the out-degree of web network follows the distribution where $\gamma_{in} \approx 2.1$, the in-degree follows the distribution where $\gamma_{out} \approx 2.4$ [59], while the autonomous system follows the distribution which $\gamma \approx 2.4$ [56]. However, some research results [60] also show that not all networks follow power law distribution, the reason can be explained by "DGX" distribution [61].

Scale-free network is the concept mentioned frequently with long tail distribution in many literatures. The definition of scale-free is: given random variable x and probability distribution $p(x)$, there is a function $g(b)$ that makes $p(bx) = g(b)p(x)$ for all x and b. In a scale-free network, the shape of distribution remains the same when scaling the range of the observed random variables. The research in [62] shows that only the network which follows power law is scale-free, while some other common distributions, for example, normal distribution doesn't have this nature.

2.3.1.2 Small Diameter

According to the definition of graph theory, if there is a path whose length is up to d between each pair of node (u, v) in graph G, the diameter of G is d. Due to the presence of outliers in the actual network, effective diameter [63] is used to denote the diameter of the network approximately. If the length between 90 % node pair (u, v) in graph G is up to d', the effective diameter of G is d'.

The study found that the Internet, Web page link network, as well as many real or Web social network all meet the characteristics of small diameter [52, 59, 64–68]. For example, the effective diameter of the MSN network is 6.6 [49], which follow the previous sociological "Six Degrees of Separation" theory.

2.3.1.3 Clustering Effect

In many real networks, the relationship between the nodes have transmissibility characteristic, i.e., if node u links to node v and node v links to node z, then u will link to z with a higher probability comparing with the randomly generated network which has the same distribution (node degree). This effect of the network is measured by clustering coefficient.

The clustering coefficient of a node is defined as follows: given node v in graph G whose degree is d, the clustering coefficient C_v of v is the ratio of the actual number of links between v's neighbors, denoted as d, to the number of links that may exist $d(d-1)/2$ [49]. Similarly, we can define C_d as the average clustering coefficient of all the nodes in graph G whose degree is d; the clustering coefficient of graph G is the mean of all nodes' clustering coefficient.

Some studies found that the clustering coefficient of actual network is higher than that in random networks with identical distribution. In addition, the clustering coefficient C_d decreases with the increases of d, and follows the power law $C_d \propto d^{-1}$ [69, 70]. Clustering effect can be used to discover the organizational structure [70–73]. Usually, the nodes with low degree in the network belong to different dense sub-graphs; the sub-graphs are connected by hubs.

2.3.1.4 Other Characteristics

Some other characteristics of real network have also been found in recent years. Some studies show that complex network is composed by the frequent appearance of basic network motifs [74, 75], while the random network with identical distribution does not have this characteristic. Recent studies also show that there is significant impact of network motifs to the transmission pattern of information [49]. The social network is also "self-healing" [59, 76], i.e., if we remove nodes randomly from the network, the connectivity of the network will not be significantly affected. But merely to remove several nodes with high degree, the connectivity of the network would be severely reduced. Self-healing is consistent with clustering effect, for sub-graphs are often connected with each other by a small number of nodes whose degree is high. Therefore, network will be divided into a number of independent sub-graphs after removing these nodes.

2.3.2 Evolution Law

Social network not only has the above static statistical properties, its evolution also follows some common patterns, including "Densification power law" and "Shrinking diameter".

With the increase of the number of nodes, the social network becomes more and more dense, i.e., the average degree of the node will increase with the increasing of nodes number. Generally, the densification of network follow the power law: $e(t) \propto n(t)^a$, where $e(t)$ and $n(t)$ represent the number of edges and nodes at time t respectively, and the range of parameter a is between 1 and 2. Research work in [77] shows that the Internet and paper citation network follow the densification power law with $a = 1.2$ and $a = 1.6$ respectively. A study of social network [78] also showed that the number of nodes (the number of users) increased rapidly in growing period, and the increasing speed was gradually slowed down in mature period.

In traditional theory, the diameter of the network is growing slowly with the increase of nodes number. However, the results in a study on the Internet and paper citation network [78] show that the effective diameter of the network decreases slowly with the increase of nodes number. Intuitive explanation for this phenomenon is that the degree of nodes increases after joint into the network, which makes the probability of finding shorter path between node pair increase accordingly.

2.3.2.1 Evolution Model

A number of explanations have been proposed on network evolution law. The Ordos model is an earlier proposed network evolution model: given n nodes, each node pair is connected by a new edge with the same probability. However, the graph generated by this model does not satisfy the long-tail distribution which is the main characteristic of real network. People also proposed the Prior Subsidiary model [53, 79] and Evolving copying model [51, 80]. In both models, each time adding a new node u to graph G, m new edges will be created. In Prior Subsidiary model, each newly created edge is connected to node v whose degree is $d(v)$ with the probability $p_u(v) \propto d(v)$. In evolving copying model, the new node is connected to m randomly selected nodes with probability β and is connected to the m neighbors of m randomly selected nodes with probability $1 - \beta$. Graph generated by Prior Subsidiary model, an devolving copying model all meet the long-tail distribution, but diameter contraction characteristic can't be established. In small-world model, a regular grid is generated first, and then the tail node of each edge is changed to another randomly selected node with probability p. A variety of networks with different characteristics can be obtained by changing the parameter p from the regular network ($p = 0$) to the random network ($p = 1$). The network generated by lower p has more partial structures constituted by short links, at the same time the diameter is longer. Higher value of p will destroy the local structure of the network, and generate shorter diameter.

Forest fire model [49] combines the advantages of the Prior Subsidiary model and Evolving copying model. It uses two parameters, the forward combustion probability p and the rearward combustion probability r, to simulate a new node "burning" the edges already existed to join the network. The experimental analysis result shows that the network generated by forest fire model, satisfies the long tail distribution and diameter contraction at the same time. The nodes with high degree act as a bridge to shorten the distance between nodes.

The above models focus on generating network which satisfies specific statistical properties, in order to explain the reasons of existing actual network. Given an actual network, how to select the model parameters in order to fit the actual statistical properties of the network? Exponential random graph [81] provides a method to select optimal model from the candidates set. But this method focuses on the measurement of the local structure in graph (e.g., which statistical properties of the node determines the generation of an edge?), and this approach is not scalable in large-scale network. The parameter estimation methods for Evolving copying model have also been proposed in [82]. But experiments show that Evolving copying model is not sufficient to simulate many actual networks. Kronecker graph model [49] use matrix Kronecker product to obtain model parameters. The complexity of the algorithm is the linear function of edge number in the network. It can effectively simulate large-scale real network, such as Epinions.com trust network.

2.3.2.2 Link Prediction

Link prediction is another important topic of network evolution law. Link prediction tries to use the intrinsic characteristics of network to model the evolution law. More specifically, it refers to "given the network snapshot at time *t*, predict the edges that will be introduces into the network since time *t* to a given future time *t*'" [83]. A common application of link prediction is to recommend friends in social network website. Another useful application is to predict the possibility of two users become coauthor. For example, when a new entry is created in the wiki site, system can recommend this record to the user who is suitable to edit it based on users' past edit history.

Formally, the input of link prediction is a relational network *G*, give a weight to each node pair <*S*, *V*> which does not exit public edge between them, and then sort by weight in ascending order, the node pairs with higher weight is the link may occur in the future. "Nearest Neighbor" and "Public Nearest Neighbor" are two typical weight calculation methods. A simple strategy of Nearest Neighbor treats the shortest path between <*S*, *V*> as the weight. It is the manifestation of network clustering effect. If node *u* links to node *v* and *v* links to node *z*, *u* may be connected to *z* with higher probability than the randomly generated network (with the same distribution). This approach does not take the number of accessible paths into account. In common sense, the more accessible paths between two nodes, the closer relationship between them. Therefore, a more reasonable method Katz [83], is to use the weighted summation of the accessible paths between all the node pair as weight. Public Nearest Neighbor method calculates the common neighbors of all nodes. The more common neighbors of node *u* and node *v*, the higher probability they are linked in the future.

A recent study [49] showed that, in MSN social network, we can do link prediction accurately by combining Public Nearest Neighbor rule and user latest activity time.

2.3.3 *Web Information Cascades*

An information cascade refers to an activity or a view point is widely adopted due to the influence of others [84]. The research in information cascade began in the field of sociology, economics and epidemiology, such as the study of Diffusion of Innovations [85] in sociology, "trend leader" in viral marketing [86], and vaccination people in epidemiological [87]. In recent years, with the development of social network, this topic also began to be concerned in the field of computer science, for example, the study of the rule of cited article in the blog [88].

The main reason that the study of web information cascade receives consideration is that, if the impact pattern between nodes in network can be decided, many decisions can be expressed as optimization problem, thereby generating direct economic benefits. Take viral marketing for example, given the marketing budget and the influence function of users in network and using a simple hill-climbing search algorithm, we can get the optimization approximate solution with a degree not less than 63 %. We can use the similar method to optimize the problem of how to place the monitors of city water pollution monitoring point.

It is different of the study in information cascade between computer science and sociology or economics. A significant difference is that the latter focuses on small-scale groups, the degree of familiarity between groups are often higher than users in the website. Some research results about blog [89–91] show that the mode of transmission that found in sociology and economics is not significant in blog. A recent study [49] found that the local structure of network has a direct impact on the dissemination of information. In addition, there are also significant differences between these local structures in different application contexts.

2.4 Trust Management

The concept of trust has been used in many research fields in computer science, including distributed authentication and authorization, trusted payment platform, and reputation system. Jøsang et al. [23] call them as trust management, in other words "The activity of creating systems and methods that allow relying parties to make assessments and decisions regarding the dependability of potential transactions involving risk, and that also allow players and system owners to increase and correctly represent the reliability of them and their systems." According to this definition, trust-based collective view prediction can be attributed to the trust management research areas, despite there are significant differences between our concerns and the three sub-areas discussed above.

Distributed authentication and authorization focus on finding a unified method to do cross-system authentication (single sign-on) and authorization in order to break the technical barriers between heterogeneous systems. The early research in this area dates back to the protocol-based system PGP public key certificate [92] and X.509 [93]. Blaze et al. [22] and others extended authentication method based on public key cryptography system to support expression and verification of security policy. In addition, the rise of automated trust negotiation [94] as well as joint authentication and access control [87, 95–97] in recent years can also be attributed to this sub-field. In these research areas, confirmation the authenticity of the identity is a major concern. Grandison [98] called it "identification trust". The identification trust is a certain relationship, the result of it is either positive or negative, and there is no intermediate state. In our study trust is an uncertain relationship, it shows the strength of trust, and it is used to generate the opinion prediction of a group.

Trusted payment platform is designed to provide users with a reliable online trading environment, to reduce the negative impact of anonymity and cross-region of Internet. Alipay in China and PayPal are the typical applications of trusted payment platform. The core idea is to increase the accountability for Internet transactions. Trusted payment platform ensure the reliability of the transactions through a series of measures. Take Alipay for example, users need to authenticate through real-name system before transaction. The real-name authentication methods include registering Alipay one-card and identifying through bank accounts. User's real identity information can be obtained through these two ways. In transaction process, users first took goods, and then pay money into his Alipay account. Sellers shipped after receipt of the notification message. Until the buyer received the goods and sent the satisfactory information to Alipay, the sellers will get the payment in their own Alipay account. When trade disputes happen, the buyers and sellers submit their evidence to Alipay and the specialized arbitration agency will do mediation. In addition, the application website of Alipay, such as Taobao, can also provide transaction commitment such as "Quality problems of seven days" to enhance the credibility of the shop.

Reputation system calculates the collecting user feedback, so that the participants in the system are able to determine whether the behavior of others follow the business rules, thereby avoiding potential risks. In addition, the reputation system motive users to participate in the collaboration by setting reasonable rules. Reputation system has been used successfully in e-commerce sites such as Taobao and eBay. Some P2P software such as eMule and Tribler [99] also use reputation systems to prevent the occurrence of the "free-rider" [100]. Reputation system has some similarities with our study, but there are significantly differences between them in research methods and purposes. Generally, the former usually takes user's behavior patterns as a starting point, trying to find a reasonable reputation function. Assume the user as a rational subject, the output of user behavior reputation function will have a positive impact, thereby reducing the possibility of bad behavior. Golbeck [17] pointed out that the standard of whether a user's behavior is good or bad in reputation system is relatively uniform. But in the group opinion prediction problem, there is no absolute standard of good and bad for a user's behavior. For the same item (such as movie), it is common for different users give opposing comments. Reference [101] proposed a transverse evaluation result on the effectiveness of the reputation system. Reference [102] classified and summarized the reputation calculation method.

2.5 Summary

In this chapter, we reviewed three research areas related to our study. The conclusion and approach in those research areas formed the foundation of our work. And we will improve the proposed models and algorithms to for usage.

Emotional analysis can convert the text comments into ratings which can reflect user opinions, so that we can focus on collective view prediction in the latter case. In the subsequent discussion of this article, we will treat comments and ratings as synonyms. In other words, we assume that any unstructured document-level review can be mapped to a real number by the sentiment analysis, or mapped to multiple real numbers through the analysis of characteristics of a class.

The output of Collaborative filtering and trust-based collective view prediction is the same. The commonly used evaluation metrics in collaborative filtering, such as prediction accuracy and coverage, can be used directly in our research. In addition, we also explore in what condition can similarity is defined as trust, and we also design a second-order Markov chain-based collaborative filtering algorithm. The algorithm is further used to combine with trust-based prediction algorithms to improve the accuracy of the prediction.

The power law distribution of the social network, small diameter and other characteristics also played a key role in our study. Small diameter feature ensures that we can find enough near neighbors to do prediction within a few steps through breadth-first search. The power law distribution characteristics determines that a little items in the Web system usually have a lot of ratings, and these popular items is the weak point to break robustness of the collaborative filtering algorithm [6]. In the trust network, a few nodes also have high-degree which impact a lot on our prediction algorithm. So when evaluating the robustness of algorithm, we must consider the impact of attacks against these high-degree nodes especially.

In the following section, we use Beta probability distribution which usually used in reputation systems research to represent the trust [103]. We also introduce two operators in belief theory [104] to do trust measurement. In addition, the results of research in the field of distributed authentication and authorization for cross-system provide a standardized solution for trust-data sharing in collective view prediction.

Chapter 3
Collaborative Filtering

With the booming of social content web sites, information overload has become a serious problem: users who face too much information cannot obtain the useful part for their own and as a result the information utilizing efficiency is low. Many existing Internet applications, such as portals, search engines and professional document indexing are aiming to help users to filter information and find the information they want. However, these tools can only meet the mainstream needs without personalized consideration. Recommender system is an important mean of information filtering and a potential method to solve the information overload problem. And collaborative filtering (CF) is the most important technique of recommender system. In this chapter, we will introduce CF method in detail, due to its successful using in business and highly correlation to our work.

To date, collaborative filtering (CF) is the best known recommendation approach for its wide use on e-commerce web site and academia. GroupLens [105], Video Recommender [106], and Ringo [107] were the first systems to user CF method to identify like-minded users and to do automate prediction. Other examples of CF systems include Amazon.com, YouTube.com, the Jester system that recommends jokes [108], and the PHOAKSA that helps users find relevant information on the WWW [109].

Algorithms for collaborative recommendation can be group into four general classes: neighborhood-base method, latent factor model, graph-based model and recommendation based on social network.

3.1 Neighborhood-Based Method

Neighborhood-based CF method is the basic algorithm in CF. It is not only studied in academia and widely used in industry. Neighborhood-based method can be categorized into two classes: user-based CF and item-based CF.

T. Luo et al., *Trust-Based Collective View Prediction*,
DOI: 10.1007/978-1-4614-7202-5_3,

3.1.1 User-Based CF

User-based CF algorithm utilizes the entire user-item data to generate predictions. It employs statistical techniques to find a set of users, known as neighbors that have a history agreeing with the target user, for example they rate different items similarly or they bought similar set of items. Once the neighborhood of users is formed, different methods can be used to combine the preferences of neighbors to product a prediction or top-N recommendations for the target user. The first step of this algorithm is to Find Neighbors and the second step is aggregation.

Many approaches have been used to compute the similarity $sim(c, c')$ between users. Commonly used techniques include:

- Correlation: In this approach [32], the Pearson correlation is used to measure the similarity:

$$sim(x,y) = \sum_{s \in S_{xy}} \left(r_{x,s} - \overline{r_x}\right)\left(r_{y,s} - \overline{r_y}\right) \Bigg/ \sqrt{\sum_{s \in S_{xy}} \left(r_{x,s}-\overline{r_x}\right)^2 \sum_{s \in S_{xy}} \left(r_{y,s}-\overline{r_y}\right)^2} \quad (3.1)$$

where S_{xy} is the set of all items correlated by both users x and y.

- Cosine-based: In this approach [2, 110], the two users x and y are treated as two vectors in m-dimensional space, where $m = |S_{xy}|$. Then the similarity between two vectors can be measured by computing the cosine of the angle between them:

$$\begin{aligned} sim(x,y) = \cos(\overrightarrow{x}, \overrightarrow{y}) &= \frac{\overrightarrow{x} \cdot \overrightarrow{y}}{||\overrightarrow{x}||_2 \times ||\overrightarrow{y}||_2} \\ &= \sum_{s \in S_{xy}} r_{x,s} r_{y,s} \Bigg/ \sqrt{\sum_{s \in S_{xy}} r_{x,s}{}^2} \sqrt{\sum_{s \in S_{xy}} r_{y,s}{}^2} \end{aligned} \quad (3.2)$$

where $\overrightarrow{x} \cdot \overrightarrow{y}$ denotes the dot-product between the vectors $\overrightarrow{x}$ and $\overrightarrow{y}$.

Various performance-improving modifications, such as default voting [2], inverse user frequency, case amplification [2], and weighted-majority prediction [111], have been proposed as extensions to these standard correlation and cosine-based techniques.

Then the value of the unknown rating $r_{c,s}$ for user c and item s is usually computed as an aggregate of ratings of some other users (usually, the N most similar) for the same items:

$$r_{c,s} = aggr r_{c',s}\left(c' \in \widehat{c}\right) \quad (3.3)$$

where $\hat{C}$ denotes the set of N users that are the most similar to user c and who have rated item s. Some examples of the aggregation functions are [3]:

$$r_{c,s} = \frac{1}{N}\sum_{c' \in \hat{C}} r_{c',s} \tag{3.4}$$

$$r_{c,s} = k\sum_{c' \in \hat{C}} sim(c,c') \times r_{c',s} \tag{3.5}$$

$$r_{c,s} = \overline{r_c} + k\sum_{c' \in \hat{C}} sim(c,c') \times (r_{c',s} - \overline{r_c}) \tag{3.6}$$

All the three kinds of the aggregation functions calculate the utility heuristically based on the users' rating and the similarity between users. Among them, the type (3.4) is the simplest form. Type (3.5) introduces the multiplier k to calculate the weighted sum, where k severs as a normalizing factor and is usually selected as $k = 1/\sum_{c' \in \hat{C}} |sim(c,c')|$. Taking into account that different users may use the rating scale differently, the adjusted weighted sum, shown in (3.6), has been widely used, where

$$\overline{r_c} = (1/|S_c|)\sum_{s \in S_c} r_{c,s} \tag{3.7}$$

where $S_c = \{s \in S | r_{c,s} \neq \emptyset\}$

User-based CF systems have been very successful in the past, but their wide-spread use has revealed some challenges such as:

- Sparsity. The count of items is large in many commercial recommender systems, such as Amazon.com recommends books and YouTube.com recommends videos. In these systems, even very active users may have only purchased or rated less than 1 % of the items. The number of ratings already obtained is usually very small compared to the number of ratings that need to be predicted. Especially with the development of e-commerce, the magnitudes of users and items grow rapidly, resulted in the extreme sparsity of users' rating data. Therefore, in such systems, user-based CF algorithm may be unable to make accuracy prediction and recommendation.
- Scalability. Traditional CF-based algorithms performance well when there are thousands of users and items in the system. But in real-life recommendation applications especially the e-commerce systems, there are millions of users and items. User-based require computation that grows with both the number of users and the number of items. With millions of users and items, it suffers from scalability problems.

The weakness of user-based CF for large and sparse datasets let researchers to explore alternative algorithm, and the item-based CF is one of the famous methods.

3.1.2 Item-Based CF

The main idea of item-based CF is to analyze the user-item representation matrix to identify relations between different items and then to user these relations to compute the prediction rating for a given user-item pair. The intuition behind this method is that a user would be interested in purchasing items that are similar to the items the user liked earlier and would tent to avoid items that are similar to the items the user did not like earlier. This method does not require identifying the neighborhood of similar users, and as a result they tend to produce much faster recommendations [110].

There are also two steps to in item-based CF algorithm. The first step is item similarity computation and the second step is prediction.

There are a number of different ways to compute the similarity between items. Here we present two such methods.

- Correlation. The correlation similarity between two items i and j is measured by computing the Pearson correlation. Let the set of users who both rated i and j are denoted by U then the correlation similarity is given by

$$sim(x,y) = \sum_{u \in U} \left(r_{u,i} - \overline{r_i}\right)\left(r_{u,j} - \overline{r_j}\right) \Big/ \sqrt{\sum_{u \in U} \left(r_{u,i} - \overline{r_i}\right)^2 \sum_{u \in U} \left(r_{u,j} - \overline{r_j}\right)^2} \quad (3.8)$$

Here $r_{u,i}$ denotes the rating of user u on item i, $\overline{r_i}$ is the average rating of the i-th item.

- Cosine-based. In this case, two items are thought of as two vectors in the m dimensional user-space. The similarity between them is measured by computing the cosine of the angle between these two vectors.

$$sim(i,j) = \cos(\overrightarrow{i}, \overrightarrow{j}) = \frac{\overrightarrow{i} \cdot \overrightarrow{j}}{||i||_2 \times ||j||_2} \quad (3.9)$$

where "." denotes the dot-product of the two vectors.

Once we isolate the set of most similar items based on the similarity measures, the next step is to look into the target user's ratings and use a technique to obtain predictions. We can compute the prediction on an item i for a user u by computing the sum of the ratings given by the users on the items similar to i [110]. Each ratings is weighted by the corresponding similarity $sim(i,j)$ between item i and j.

$$P_{u,i} = \frac{\sum_{\text{all similar items, N}} \left(s_{i,N}^{*} r_{u,N}\right)}{\sum_{\text{all similar items, N}} \left(|s_{i,N}|\right)} \quad (3.10)$$

The aggregation methods introduced in user-based CF can also be used here. And regression method can be used to get the prediction. Some results show that item-based CF method hold the promise of allowing CF-based algorithms to scale to large data sets ad at the same time produce high-quality recommendations. And item-based CF has been widely used in commercial web sites, such as Amazon.com and Netflix that recommends movie DVD.

3.1.3 Comprehensive Analysis of User-Based CF and Item-Based CF

User-based CF recommends the items which are interested by the users who are similar to the target user. Item-based CF recommends the items which are similar to the items ever rated or purchased by the target user. The recommendations of user-based CF are more socialization. User-based CF is focus on the popularity items in small interest groups. The recommendations of item-based CF are more personalized and reflect the user's interest history.

The user's interest is not particularly refined in News website. The vast majority of users prefer to read popular news. Although some users like sports news and some prefer entertainment news, the granularity if fairly thick. It would not happen that a user only likes reading a specific topic of news. Therefore, the recommendations of new pays more attention to socialization and personalization is second. So it is suitable to use user-based CF in news recommendation applications. Digg.com use user-based CF to do news recommend.

But in books, electronic commerce, as well as film website, the user interest is fixed and lasting. A researcher purchase academic books, he doesn't need reference popularity and only need to make judgment based on his domain knowledge. As a result item-based CF is widely used in Amazon, Netflix and other commercial websites. In addition the item update speed is not particularly fast in such systems, therefore update item similarity matrix once a day is acceptable.

User-based CF needs to maintain a user similarity matrix, and item-based CF needs to maintain item similarity matrix. If the number of user is large, it is consuming to compute and maintain user similarity matrix. Similarly, if the count of item is large, it is consuming to maintain item similarity matrix.

The comprehensive analysis of user-based CF and item-based CF is shown in the following Table 3.1).

3.2 Latent Factor Model

Latent Factor Model relies on a statistical modeling technique that introduces latent class variables in a mixture model setting to discover user communities and prototypical interest profiles.

Table 3.1 The comprehensive analysis of user-based CF and item-based CF

	User-based CF	Item-based CF
Scalability	Suitable for less user situation. If the number of user is large, the calculation of user similarity matrix is time consuming	Suitable for the situation that the number of items are obviously small than the number of users. If the number of item is large (i.e. web pages), the calculation of item similarity matrix is time consuming
Domain	Suitable for domain which has strong timeliness and user personalized interest is not obvious	Suitable for domain which needs high level user personalization, and have many long tail items
Dynamics	It is not always change the recommendations immediately if user has new behavior	It will certainly lead to the real-time change of recommend results when new user behavior happens
New user problem	Cannot provide recommendation for new users, for the user similarity matrix is computed offline and update every once in a while	Can provide recommendations to new user just after he/she rated one item
New item problem	If there are users rated new items, the system can recommend it to the similar users	Cannot recommendation new items to users immediately, for the item similarity matrix is computed offline and update every once in a while
Interpretability of recommendations	Hard to provide convincing explanations of recommendation	Be able to provide convincing explanations based on users' historical behavior

3.2.1 Singular Value Decomposition

We will focus on models that are induced by Singular Value Decomposition (SVD) on the user-item ratings matrix. Recently, SVD [12, 112] models have gained popularity due to their attractive accuracy and salability. Given a rating matrix $R \in \Re^{m \times n}$, where $R[u][i] = r_{ui}$, the main idea of SVD models is to break R into the multiplication of three matrices, which are $U \in \Re^{k*n}$, $V \in \Re^{k*m}$, $S \in \Re^{k*k}$. S is a diagonal matrix, and the elements on its diagonal is the characteristic values of R. Select the largest k characteristic values to form the diagonal matrix S_k, and at the same time select the corresponding row vectors and column vectors of U and V to form U_k and V_k. $\hat{R} = U_k^T \cdot S_k \cdot V_k$. $\hat{R}$ is the final complemented matrix of the rating matrix R. And $\hat{R}(u,i)$ is the predicted value of user u to item i.

3.2.2 Regularized Singular Value Decomposition

However, applying SVD directly is difficult due to the high portion of missing rating. Previous approaches [112] rely on imputation to fill in missing ratings and make the rating matrix denser. However, imputation can by very expensive as it significantly increases the amount of data. In addition, the data may be considerably dirty and distorted due to the inaccurate imputation. Hence, more recent work [113] suggested modeling directly on the observed rating, and to avoid overfitting problem through a regularized model. In such model, the prediction of user u's preference on item I is made by:

$$\hat{r}(u,i) = \bar{r} + b_u + b_i + p_u^T \cdot q_i \tag{3.11}$$

where $\bar{r}$ is the average rating of all the knowing ratings. b_u is user bias, used to describe a user tends to rate their favorite things or tend to rating things they do not like. b_i is item bias, used to describe the rating of an item higher or lower than the average rating. p_u is the user- factor vector and q_i is the item-factor vector. The model is trained by minimizing the following cost function on observed ratings:

$$\min \sum_{(u,i)\in D} \left(r_{ui} - \bar{r} - b_u - b_i - p_u^T \cdot q_i\right)^2 + \lambda\left(b_u^2 + b_i^2 + \|p_u\|^2 + \|q_i\|^2\right) \tag{3.12}$$

where D is the observed rating set, is a regularization parameter to avoid overfitting.

The optimization problem of the above parameters can be solved by simple gradient descent method [114]. First take partial derivative p_u and q_i:

$$\frac{\partial C}{\partial p_u} = -2q_i e_{ui} + 2\lambda p_u$$

$$\frac{\partial C}{\partial q_i} = -2p_u e_{ui} + 2\lambda q_i \tag{3.13}$$

where $e_{ui} = r_{ui} - p_u^T \cdot q_i$. And then using stochastic gradient descent method to update and iteratively,

$$p_u \leftarrow p_u + \alpha(e_{ui} q_i - \lambda p_u)$$

$$q_i \leftarrow q_i + \alpha(e_{ui} p_u - \lambda q_i) \tag{3.14}$$

where α is the learning rate.

Latent factor models are based on machine learning algorithms and have good theory basis. And Latent factor models comprise an alternative approach to Collaborative Filtering with the more holistic goal to uncover latent features that explain observed ratings.

3.3 Graph-Based Collaborative Filtering

Graph-based CF methods model the interaction between users and items on a graph. The users and items are represented as nodes and the relationship between them are indicated as edge. If a particular user have purchased or viewed an item, there is a line between them. The weight of the line is determined by the application and the algorithm used. Then the similarity of the nodes is computed from a global perspective, instead of local pair-based computation of neighborhood.

3.3.1 Bipartite Graph Model of Collaborative Filtering

The data form in graph based recommendation is $G(U, I, E, w)$, where U is the set of user nodes, I denotes the set of item nodes. If user u had rated or browsed item i, there is an edge $e(v_u, v_i) \in E$ links the user node v_u and item node v_i. And w is the weight of edge. The left part in Fig. 3.1 denotes the user-item relationship. For example, user A interacted with item a and c, and user C interacted with item b and d. The right part of the figure is the bipartite graph built based on the relationship. For example, A interacted with a and c, so there are edges $A \rightarrow a$ and $A \rightarrow c$.

The task of recommendation is to measure the interest of user to item. In bipartite graph model the task is transferred to measure the distance of user node to item node. In Fig. 3.1 there isn't direct path between A and d, but A and d can connected by two 3-length paths, which are $A \rightarrow a \rightarrow B \rightarrow d$ and $A \rightarrow c \rightarrow B \rightarrow d$. So we guess user A may have interest to item d.

3.3.2 Graph-Based algorithm of Collaborative Filtering

In graph-based algorithm of CF, the recommendation task can be transformed to measure the relevance between the user vertice v_u and the item vertices which are not connected directly with v_u. The higher the correlation of the item, the higher weight it is in the recommendation list. In common, the relevance between vertices is determined by the following three elements:

- The number of paths between two vertices.

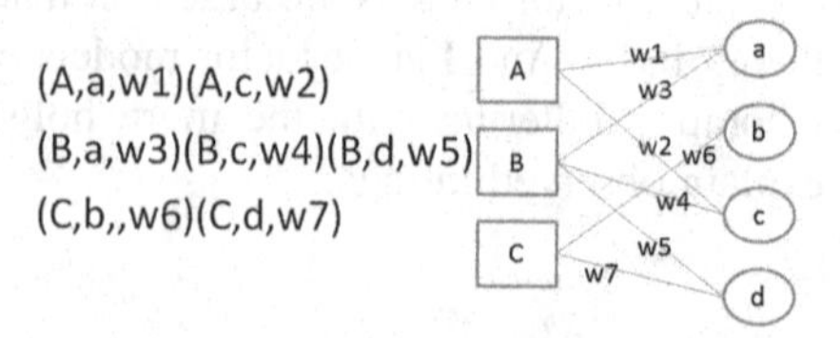

Fig. 3.1 Bipartite graph of the user-item relationship

- The length of paths between two vertices.
- The nodes which are in the path between two vertices.

The common characteristics of a pair of vertices with high correlation are:

- There are many paths between the two vertices.
- The length of paths between them is relative short.
- There are no vertices whose out-degree is high in the paths between the two vertices.

Based on the above main points, there are many techniques to measure the relevance between two vertices in graph [115]. Here we introduce a PersonalRank algorithm [116] which based on random walk theory.

If we want to provide recommendation list for user u, we can random walk from the corresponding vertices v_u in user-item bipartite graph. When walk to a node, first decide whether to continue walk or stop this walk and began random walk again from v_u in accordance with the probability α. If decide to continue walk from the current node, then select a node from the connected nodes of the current node in accordance with the uniform distribution. After a number of random walks, the visited probability of each node will converge to a number. The weight of the item in the final recommendation list is determined by the above visited probability. The corresponding formula is showed as follows:

$$Rank(v) = \begin{cases} \alpha \sum\limits_{v' \in in(v)} \frac{Rank(v')}{|out(v')|} & (v \neq v') \\ (1-\alpha) + \alpha \sum\limits_{v' \in in(v)} \frac{Rank(v')}{|out(v')|} & (v = v') \end{cases} \quad (3.15)$$

where $in(v)$ is the set of nodes which are connected with v, and $|out(v')|$ is the out-degree of node v'.

3.4 Socialization Collaborative Filtering

Socialization recommendation is a reproduction of what happened in social reality. In real-life, people often make decision on friends' selection. Recommendations from personal acquaintances or opinions posted by consumers online are the most trusted forms of advertising, according to the latest Nielsen Global Online Consumer Survey of over 25, 000 Internet consumers from 50 countries. And 90 % or consumers surveyed noted that they trust recommendations from people they know, while 70 % trusted consumer opinions posted online [117]. It can be concluded from the survey that it is important to add the trust degree of the recommended items to users by friends' selection. In this subsection we will introduce the way to get socialization data, neighborhood-based socialization recommendation algorithm, and graph-based socialization recommendation method.

3.4.1 Gathering Socialization Data

The common ways to get socialization data is showed as follows:

3.4.1.1 User Registration Information

Some websites ask users to provides some personal details, such as school, company information when registration. We can analysis which group of people is working in the same company or which group of people has studied in the same university based on that information. Therefore, user registration information is a kind of implicit socialization data.

3.4.1.2 User Location Information

We can get user's IP information if the user access to Internet via desktop or laptop. And we can get more accurate GPS information if the user access to Internet via mobile phone. User location information describes users' social relationship in certain extent. We can find out which group of people is in the same city, in the same school or even in the same office block. So it is reasonable to assume people in the same dominate building or in the same office block may have friendship, and people in the same restaurant may have similar food preferences.

3.4.1.3 Forum

Forum is the typical product of Web 1.0. A group of users can make discussion on a topic in forum. A specific group contains people who have similar taste. A user can add to some different groups. It is can be concluded that people in the same group may have friendship or have similar interest.

3.4.1.4 Instant Message Tool

User can communicate by text, voice, video and other ways through instant message tool. User will have a list of contacts, and he/she often divide contacts into group. By analyzing the list and group information, we can get the user's social network relations. And we can find out the familiarity of the contact with current user by measuring the frequency of the chats between them.

But instant message tool is a close system. Get the instant message of user is very difficult and involve privacy issues. In common, most of the users are not willing to public their contacts list and instant message history.

3.4.1.5 Social Network

It is difficult to obtain socialization data because of privacy issues for some of methods of the above ways. Although some information is readily available, they are implicit socialization data and it is difficult to infer social relationship between users. The emergence of social network websites, such as Facebook and Twitter, break the above bottleneck. In social network website, user can build a public personal homepage and public his friend list. Users can share pictures, music, and videos or discuss hot news. Personalized recommendation system can take advantage of the social relationship data and user behavior history data which are policed by social network website to assist user solve the information overload problem and find out friends and contents user may interested in.

3.4.2 Neighborhood-Based Socialization Recommendation Algorithm

The input of this algorithm includes social network data and user behavior history data. The social network data provide the friendship between users. And the user behavior data provide the user's interest information. The simplest method is to recommend a collection of items which are liked by the user's friends. The interest degree p_{ui} of user u to item i can be computed by following formula:

$$p_{ui} = \sum_{v \in out(u)} r_{vi} \tag{3.16}$$

where $out(u)$ is the friends collection of user u. If user v like item i, then $r_{vi} = 1$, otherwise $r_{vi} = 0$. The familiarity and similarity of different friends with the current user is different. So we should take that difference into account when doing recommendation:

$$p_{ui} = \sum_{v \in out(u)} w_{uv} r_{ui} \tag{3.17}$$

where the weight w_{uv} is includes two parts, *familiarity* and *similarity*. The *familiarity* between user u and user v is defined as the familiarity of the two users in real-life. In common, people are willing to take the familiar friends' recommendations. Familiarity can be measured by the ratio of common friends between users:

$$familarity(u, v) = \frac{|out(u) \cap out(v)|}{|out(u) \cup out(v)|} \tag{3.18}$$

Similarity describes the degree of similarity of interests between two users. We are familiarity with our parents, but we have different preference with them. So similarity should be taken into account in socialization recommendation algorithm. *Similarity* can be measured by the coincidence level of the common liked items:

$$similarity(u,v) = \frac{|N(u) \cap N(v)|}{|N(u) \cup N(v)|} \tag{3.19}$$

where $N(u)$ is the collection of items which are liked by user u. The weight of *familiarity* and *similarity* should be determined by the characteristic of application. For example if the system designer wants to pay much attention to familiarity between users, the weight of *familiarity* can be increased, and at the same time the weight of *similarity* should be decreased accordingly.

3.4.3 Graph-Based Socialization Recommendation Algorithm

The advantage of graph model is that it can describe a variety of data and relationships on a figure. User's social network can be represented as social networking graph. The behavior of users to items can be represented as a bipartite graph. And those two kinds of graphs can be combined into one graph. If any behavior happened of a user to an item, there will be an edge between them. If two users have friendship, there also will be an edge between them.

Figure 3.2 is a typical example of the combined graph. Uppercase letter node indicates user, and lowercase letter indicates item. User A have rated or purchased on item a, b, and f. And user A and B, C are friends.

The weight of the edge between user and user node can be measured by familiarity and similarity discussed in the above sub section. The weight of the edge between user and item node can be by measured by the degree of likeability of user to item. The weight of the two kinds of edges should be determined by the

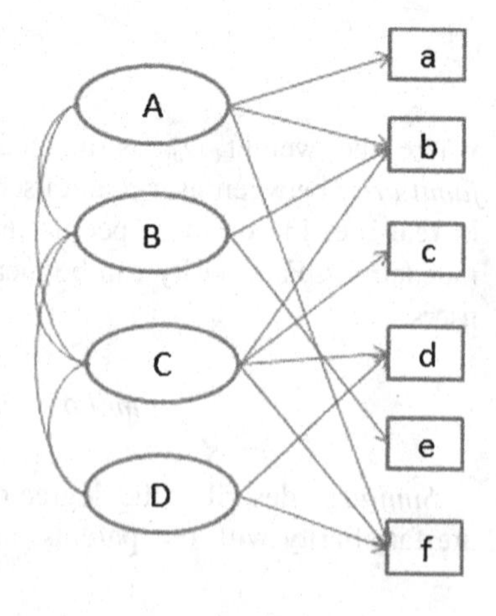

Fig. 3.2 Combine graph of user social network and user-item relationships

characteristic of application. For example if the system designer wants to pay much attention to social relations between users, the weight of familiarity and similarity can be increased. After defining the node, edge and weight of edge, graph-based recommendation algorithm, such as PersonalRank method can be used to produce socialization recommendations.

In some social networks, there are also group relationships between users. There may not be friendship between two users, but they can be belonged to the same group, for example the group often discussed on English cultural. Figure 3.3 is an example of the above situation. In this graph, there is not only friendship between users, but also membership. We can use the combination of membership and friendship to define the weight of edge between users. After defining the node, edge and weight of edge, graph-based recommendation algorithm, such as PersonalRank method can be also used to produce socialization recommendations [113].

There are also other branches of collaborative filtering, for example [118] found out that there are many differences between academic recommendation and commercial one. Ref. [119] proposed a method which combines content correlation and collaborative filtering approach to do academic recommendation. Customer preferences for products are drifting over time. Product perception and popularity are constantly changing. Similarly, uses inclination are evolving. Thus, modeling temporal dynamics should be a problem when designing recommender systems and algorithms. References [120, 121] both proposed a time-weighted item based collaborative filtering method by reducing the influence of older data when predicting user's further behavior. The basic idea of those approaches is that the user's further decision is depend on his/her recently selection. Similarly, user based collaborative filtering algorithm can be extend to support temporal recommendation. The key idea is the similar user's recently decision has a great effect on his/her further choices. Ref. [122] treated CF as a time-dependent iterative prediction problem, and proposed an algorithm to assign and update per-user neighborhood sized automatically other than setting global parameter. Ref. [123] made prediction of movie ratings for Netflix by modeling the temporal dynamics based on a factorization model. Ref. [124] proposed a graph-based method to do academic recommendation. This method can capture learner's preferences and learning context accurately and dynamically.

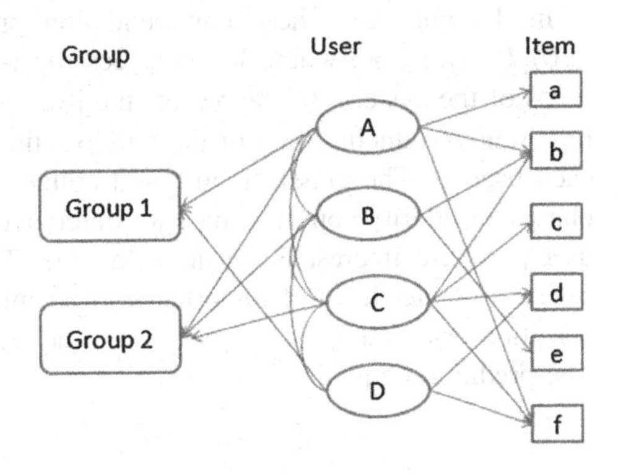

Fig. 3.3 Combine graph of user friendship, user membership and user-item relation

3.5 Dynamic Model in Collaborative Filtering

In recent years, time information is more and more important in collaborative filtering (CF) based recommendation system because many systems have collected rating or log data for a long time, and time effects in user preference is much stronger. Ref. [178] pointed out that time information influences CF from four different ways.

Algorithms for using time effects in collaborative filtering can be group into three general classes: dynamic neighborhood-based model, dynamic latent factor model, dynamic graph-based model. In the following part, we will introduce the first two models in detail and give a case study on dynamic graph-based method.

3.5.1 Dynamic Neighborhood-Based Model

Dynamic neighborhood-based CF model proposed a time-weighted item based collaborative filtering method by reducing the influence of older data when predicting user's further behavior. The basic idea of those approaches is that the user's further decision is depend on his/her recently selection. Similarly, user based collaborative filtering algorithm can be extend to support temporal recommendation. The key idea is the similar user's recently decision has a great effect on his/her further choices.

Take Item-KNN which discussed in Sect. 3.1 for example. The item-based approaches for collaborative filtering identify the similarity between two items by comparing users' ratings on them. In traditional model, ratings produces at different times are weighted equally. But in dynamic Item-KNN model [178, 121], the time weights for different items are computed in a manner that will assign a decreasing weights for older data.

The main idea of neighborhood-based time weight algorithm [120] is to find appropriate time weights for items in order that the items rated recently are able to contribute more to prediction of the recommendation items. More recent data should have higher value in the time weighting.

In dynamic CF, the recommendation problem can be defined as a tuple $<U_i, I_j, O_{ij}, T_{ij}>$, where U_i identifies the i-th user of the system, I_j is the j-th items of the system, O_{ij} represents the i-th user's opinion on the j-th item and T_{ij} represents producing time of the opinion, find a list of k recommended items for each user U. The classic item-based collaborative filtering algorithms have two phases: similarity computation and preference prediction. As assumed before, the user purchase interest is sensitive to time. The recommendation process should assign a greater level of importance to recent data. So in the phase of preference perdition, each rating is assigned a weight defined by a function f(t) to the time t. The prediction equation is:

$$O_{ij} = \frac{\sum_{c=1}^{k} O_{ic} \times sim(I_j, I_c) \times f(t_{ic})}{\sum_{c=1}^{k} sim(I_j, I_c) \times f(t_{ic})} \tag{3.20}$$

where I_c identifies the nearest neighbors of the j-th item, O_{ij} represents the i-th user's opinion on item j-th item and t_{ic} represents the time the user's opinion O_{ic} was produces.

The time function f(t) is a monotonic decreasing function, which reduces uniformly with time t and the value of the time weight lies in the range (0, 1). In other words, all the data contribute to the recommendation items, while the most recent data contributes the most. The old data reflects users' previous preferences. It should have small weight in the prediction of recommendation. In [120], they choose an exponential form for the time function to achieve the goal. The exponential time function is widely used in many applications in which it is desirable to gradually decay the history of past behavior as time goes by. Firstly, it defines a half-life parameter T_0 as:

$$F(T_0) = \frac{1}{2} f(0) \tag{3.21}$$

That is to say the weight reduces by 1/2 in T_0 days.

Then they define the decay rate λ as:

$$\lambda = \frac{1}{T_0} \tag{3.22}$$

The time function is as follows:

$$f(t) = e^{-\lambda t} \tag{3.23}$$

From the above equation, we can observe that the value of the time function is in the rage (0, 1), and it reduces with time. The more recent the data, the higher the value of the time function is. The exponential function satisfies the dynamic recommendation's needs well. There are still other time functions, for example, logistic function. The chosen of time decay function is depends on the application and environment.

3.5.2 Dynamic Latent Factor Model

Factorized models, also known as latent factor models and latent class models are widely used in collaborative filtering. As we introduced in Sect. 3.2, the prediction of user the prediction of user u's preference on item i is made by:

$$\hat{r}(u,i) = \bar{r} + b_u + b_i + p_u^T \cdot q_i \tag{3.24}$$

where $\bar{r}$ is the average rating of all the knowing ratings. b_u is user bias, used to describe a user tends to rate their favorite things or tend to rating things they do not like. b_i is item bias, used to describe the rating of an item higher or lower than the average rating. $p_u$is the user-factor vector and q_i is the item-factor vector. The model is trained by minimizing the following cost function on observed ratings:

$$\min \sum_{(u,i)\in D} \left(r_{ui} - \bar{r} - b_u - b_i - p_u^T \cdot q_i\right)^2 + \lambda\left(b_u^2 + b_i^2 + \|p_u\|^2 + \|q_i\|^2\right) \quad (3.25)$$

Dynamic latent factor model is adding time weight in the traditional recommendation methods. Ref. [178] proposed that there are four main types of time effects in collaborative filtering:

Firstly, the interest of whole society changes with time.

Secondly, rating habit of users changes with time. For example, a man may firstly give 5 stars to those items he likes, but after a period of time, he will give no more than 4 stars to those items he enjoys. This means, he is pickier when time goes on.

Thirdly, the items' popularity changes with time. A movie may lose popularity because it is too old or gets popularity because it wins some awards or its actor becomes popular.

The last time effect is, users may change their preferences with time. Many events can cause a user changes his/her preferences. For example, a boy likes watching cartoon when he is young, but he enjoys war films when he grows up.

And they add these four time effects in the above factorized model. In the following, we will introduce the four factors and the final model in detail.

3.5.2.1 Time Bias

The interest and habit of the whole society change with time. At different ages, peoples enjoy different things. In a recommender system, average rating of all items changes with time. Take Netflix data (http://www.Netflixprize.com) as example, the average rating varies between 3.25 and 3.5 before March 2004 and it varies between 3.55 and 3.75 after Mach 2004. Many reasons can cause this effect. For example, movies quality is improved after March 2004 or people like those movies released after March 2004. This effect is called time bias, it is used by adding a scalar b_t to Eq. 3.24:

$$\hat{r}(u,i) = \bar{r} + b_u + b_i + b_t + p_u^T \cdot q_i \quad (3.26)$$

where $t = t_{ui}$.

3.5.2.2 User Bias Shifting

Users may change their rating habit with time. For example, a user may rate items they like in a period of time and rate items they dislike in another period of time. Furthermore, some users may firstly tend to give no more than 4 stars to the items they enjoy, but after a period of time, they tend to give 5 stars to the items they like. This time effect is called user bias shifting. In order to model this time effect, the user bias vector b_u in Eq. 3.24 is replaced by a time-dependent function $b_{u\tau}$:

$$b_{u\tau} = b_u + x_u^T z_\tau \tag{3.27}$$

where $\tau = \tau_{ui} = t_{ui} - t_u$ is the number of days after user u enters the recommender system and $x_u, z_u \in \Re^f$ are two latent factor vectors for user u and time τ. In this way, the model becomes:

$$\hat{r}(u,i) = \overline{r} + b_u + b_i + b_t + p_u^T \cdot q_i + x_u^T z_\tau \tag{3.28}$$

3.5.2.3 Item Biases Shifting

The popularity of items is changing with time. Several events can cause an item to become more or less favorable. For example, if an actor wins Oscar's best actor award, his previous movies may become more favorable. However, movies will loss popularity when time goes on generally. In this way, the item bias b_i is also a function that changes with time. It is easier to capture the time effect in item bias by using a time-dependent item bias model:

$$b_{iw} = b_i + s_i^T y_w \tag{3.29}$$

where $w = w_{ui} = t_{ui} - t_i$ is the number of days after item i's first rating was assigned. $s_i, y_w \in \Re^f$ are two latent factor vectors for item i ant time w. Here, $s_i^T y_w$ represents the fluctuation of item i's popularity with time w. In this way, the model becomes:

$$\hat{r}(u,i) = \overline{r} + b_u + b_i + b_t + p_u^T \cdot q_i + x_u^T z_\tau + s_i^T y_w \tag{3.30}$$

3.5.2.4 User Preference Shifting

Users change their preferences with time. For example, a boy likes cartoon when he is young and rates a cartoon "Toy Story", released in 1995, the highest score 5 stars in 1998. However, when he grows up, he does not like cartoon and rates another cartoon "The Incredibles", released in 2004, the low score 2 stars in 2005. Furthermore, users may change their attitude toward actors and directors. This time effect is called user preference shifting. In the traditional model, $p_u^T \cdot q_i$ represents the preference of user u on item i. In the dynamic model, $p_u^T \cdot q_i$ is replaced by a time related model:

$$preference(u,i) = p_u^T \cdot q_i + \sum_{k=1}^{f} g_{u,k} \cdot l_{i,k} \cdot h_{\tau,k} \quad (3.31)$$

where $g_u, l_i, h_\tau \in \Re^f$ are three latent factors for user u, item i and time τ. So the final model becomes:

$$\hat{r}(u,i) = \bar{r} + b_u + b_i + b_t + p_u^T \cdot q_i + x_u^T z_\tau + s_i^T y_w + \sum_k g_{u,k} \cdot l_{i,k} \cdot h_{\tau,k} \quad (3.32)$$

This dynamic model is trained by minimizing the following regularized cost function:

$$\min \sum_{(u,i)\in D} \begin{array}{l} (r_{ui} - \hat{r}(u,i))^2 + \lambda(b_u^2 + b_i^2 + b_t^2 + \|p_u\|^2 + \|q_i\|^2 + \|x_u\|^2 \\ + \|z_\tau\| + \|x_i\| + \|y_w\| + \|g_u\| + \|l_i\| + \|h_\tau\|) \end{array} \quad (3.33)$$

where λ is regularization parameter which is chosen by cross-validation. In order to solve this optimization problem, a simple gradient descent method which is introduced in Sect. 3.2 is applied.

3.5.3 *Case Study: Dynamic Graph-Based Collaborative Filtering*

3.5.3.1 Introduction

Ref. [118] pointed out that making recommendations to learners in academic context differ from recommending in commercial domains, in which only interest matter. A learner's evaluation about learning material depends not only on how interesting it is, but also on the degree that the material helps them meet their cognitive goals. Learner preferences for academic material are drifting over time. Resources perception and popularity are constantly changing as new academic trend emerges. Leaner's cognitive level also plays an important role, for freshman likes basic items and senior users prefer advanced ones.

To date, collaborative filtering (CF) is the best known recommendation approach for its wide use on e-commerce web site. In this technique, ratings produced at different times are weighted equally. That is to say, change in user's interest and context is not taken into consideration. Recently, Koren [123] modeled the time factors for each user separately in a factorization model on the explicit rating data of Netflix. But the above framework is not able to deal with implicit data. Moreover, explicit rating data is not always available, and the practical recommendation system has to model the implicit feedback, including, learner's browsing history, cite history, etc. Therefore, it is a great challenge to model learners' preference and academic trend over time accurately on implicit data.

In this case study, we model the drifting of learner's preference and academic trend over time by Dynamic Transfer Chain (DTC) [119]. The chain introduces four elements which present the time factor in learning context, including user temporal weight, item durability, item overall popularity, and item current popularity. Then we propose a novel algorithm Dynamic Academic Recommendation on Graph (DARG) to make accurate top-N recommendations on implicit data. We systematically compare our method with other start-of-the-art algorithms on an open dataset. The results confirm that our approach can incorporate temporal information effectively.

3.5.3.2 Related Work

(1) Academic Recommendation

Many work such as [179, 180] has been done in E-learning Personalization field. But these techniques are focused on intelligent tutoring system, not the academic recommendation to support lifelong learning. Most current recommendation techniques that have been used in e-learning were established in the same way as in e-commerce without taking into account the pedagogic theory. Some research projects in this area are based on different kind of collaborative filtering method, such as the I-Help system [181], and the Context e-learning with Broadband Techniques [182]. Ref. [183] realized the specific characteristics of academic recommendation, but unfortunately the model they explored are still following the traditional ways such as collaborative filtering, without the introduction of pedagogic factors.

There are also some literatures studied on research paper recommendation. Ref. [184] proposed an academic paper recommender system to provide related papers to users based on text similarity. Ref. [185] presented a paper recommendation system which made use of not only keyword-based search but also user's explicit and implicit feedback. Ref. [190] introduced a research paper recommending algorithm based on the citation graph and random walk model. However, all those work are based on keywords to a certain extent. Using keyword to do recommendation may lead to overspecialization problem, for the user is limited be being recommended items that are similar to those already rated [2]. In addition, there are many types of academic resources, including course, conference information, and blogs and so on, rather than just research papers

(2) Graph-based Recommendation

Graph-based methods model the interaction between users and items on a graph. The users and items are represented as nodes and the relationship between them are indicated as edge. If a particular user have purchased or viewed an item, there is a line between them. The weight of the line is determined by the application and the algorithm used. Then the similarity of the nodes is computed from a global perspective, instead of local pair-based computation of neighborhood.

References [186, 187] proposed a similarity measure between nodes of a graph integrating indirect paths, based on the matrix-forest theorem. Ref. [188] adopted the label propagation method in semi-supervised learning based on a video co-view graph to make video recommendation for YouTube. Recently some authors considered similarity measures based on random-walk model. Ref. [190] proposed a random walk based scoring algorithm to rank products according to expected user profiles. Ref. [189] presented a two-layered graph model for products recommendation based on the association strengths between a customer and products. However, all these work do not consider the temporal information in the graph model.

(3) Temporal Dynamics in Recommendation

References [119, 121] both proposed a time-weighted item based collaborative filtering method by reducing the influence of older data when predicting user's further behavior. The basic idea of those approaches is that the user's further decision is depend on his/her recently selection. Similarly, user based collaborative filtering algorithm can be extend to support temporal recommendation. The key idea is the similar user's recently decision has a great effect on his/her further choices. All the above methods have a disadvantage due to latest data are not always reliable while sometimes the older data are important. Ref. [192] treated CF as a time-dependent iterative prediction problem, and proposed an algorithm to assign and update per-user neighborhood sized automatically other than setting global parameter. Ref. [123] made prediction of movie ratings for Netflix by modeling the temporal dynamics based on a factorization model. However, all those methods are focused on explicit data instead of implicit feedback.

In a word, to the best of our knowledge, there has not been any study on modeling learner's preferences and academic trend based on implicit feedback and making use of the model to do recommendation. Our focus is to present the drifting of learner's profile and academic context over time on graph, and provides high quality of top-N recommendations.

3.5.3.3 Methodology

In this section, we review the bipartite graph model of CF first, and then we illustrate how to model learner's preferences and academic trend by Dynamic Transfer Chain (DTC). And we will present the novel algorithm called Dynamic Academic Recommendation on Graph (DARG) to make temporal dynamic top-N recommendation.

In our model we find the shortest paths between user u and item i first. And the degree of interest of u to i is determined by DTC.

(1) Dynamic Transfer Chain (DTC)

The Dynamic Transfer Chain (DTC) contains four elements which can by classified into two categories. The first category is about the academic resources, including item durability; item overall popularity, and item current popularity. The other is concern on the learners that is the user temporal weight. In this part, we will introduce them in detail.

- Item Durability:

$$w_d = \frac{MaxT_i - MinT_i}{Duration_i} \tag{3.34}$$

$MaxT_i$ is the last time the item i is cited or bowered, and $MinT_i$ represents the earliest time correspondingly. $Duration_i$ is the time span from when the item i join the website to the present, say three months, four years or even much longer. This parameter represents whether the academic resource can withstand the test of time. High quality research items can be popular for a long time, in contrast to some items only like a flash. If an academic resource got attention from its appearance to now, the weight is 1. The smaller the popular time span accounting for the proportion of its duration, the closer the value of weight is to 0.

- Item Overall Popularity:

$$w_{op} = e^{-\frac{MaxP - P_i}{MaxP - MinP}} \tag{3.35}$$

MaxP is the popularity of the most popular item in the system. *MinP* is the popularity of the most unwelcome item. Popularity is defined as the number of clicks of the item. The popularity of the current item is denoted as P_i. This parameter measures the popularity of an item totally. If i is the most popular item in the system, the weight is $e^0 = 1$. Contrastingly, if i is the most unwelcome item, the weight is $e^{-1} = 0.368$.

- Item Current Popularity: To measure whether the item is out of style.

$$w_{cp} = e^{-\frac{MaxT - MaxT_i}{TotalDuration}} \tag{3.36}$$

MaxT represents the last rating or citing time in the system, while $MaxT_i$ is the last rating or citing time of the item i. *Total Duration* is the time span of the whole system, from the first time the item in the system had been rated or cited to the last time the rating or citing happened. This parameter denotes whether the current item is still popular.

Item Overall Popularity denotes the simple popularity of an item, while *Item Durability* and *Item Current Popularity* can catch the academic trend. We can measure an academic resource comprehensively even through linear sum of the above three parameters. The sum of weight is high for the classic item, which is popular for a long time. And the new item in the system is also high, for it may present the trend of a particular field, while the results will be low for the flash or the low quality items.

- User Temporal Weight:

$$w_{uti} = e^{-\frac{\varphi(MaxT_u - T_{ui})}{MaxT_u - MinT_u}} \tag{3.37}$$

$MaxT_u$ is the last time of user u's rating or citing behavior happened. $MinT_u$ represents the first time user u's behavior occurred. T_{ui} is the time the user u rating or citing item i. Learner's recently behavior have more impact to recommendation, while his/her early usage data have less impact, φ is the parameter to adjust the weight.

We also find that the interest of the freshmen is changing rapidly, since them may have not been found their real focus. However, the advanced ones' focus will not be changed easily because they know what he needs clearly. In this study, we use the time span of the user's using system to model the cognitive level of users simply. If the user has been using the system for a long time, he/she is divided to the advanced user category, while the man/woman whose using history is short belongs to the non-advance category. While this classification way is rough, the experiment results show an improvement can be achieved by this change.

$$w_{uti} = \begin{cases} e^{-\frac{\varphi_1(MaxT_u - T_{ui})}{MaxT_u - MinT_u}} if(MaxT_u - MinT_u) < \sigma \\ e^{-\frac{\varphi_2(MaxT_u - T_{ui})}{MaxT_u - MinT_u}} if(MaxT_u - MinT_u) \geq \sigma \end{cases} \tag{3.38}$$

σ is the threshold of weeks that user u entering the system. Parameter φ_1 and φ_2 adjust the weight. If φ_1 equals 1, the weight of the latest rating is $e^0 = 1$, and the weight of the earliest rating is $e^{-1} = 0.368$. If φ_2 equals 2, the two weight are $e^0 = 1$ and $e^{-2} = 0.135$ correspondingly. The decay of the weight over time can be adjusted by different value of σ, φ_1, and φ_2.

(2) Dynamic Transfer Chain based Graph (DARG)

In order to make good recommendation based on graph, we summarize three rules that if user node u gives high preference to item node i,

- Node u and node i is connected.
- There is a short path between node u and node i.
- The weight of the short path is high based on DTC.

The process of the DARG can be summarized as the following four steps:

- Find the shortest paths between user u and item i based on breadth-first-search.
- Calculate the weights of the shortest paths based on DTC.
- The predicted preference of user u to item i is the linear sum of the above weights.
- Sort the preferences and find the top N items as the recommendations.

The pseudo code of DARG to make top-N recommendations for active user u to item i at time t is showed in Table 3.2.

3.5.3.4 Experiments

In this section, we will present the dataset we adopt, the evaluation metric, the compared methods, the parameter setting, and the overall accuracy comparison.

(1) CiteULike Dataset

CiteULike [193] provides free online service to manage academic publications. It allows users to tag or bookmark academic resources. CiteULike dataset can be downloaded from [194]. The latest "who bookmark what" data released by CiteULike contains 113,567 user, 3,678,681 research papers, 16,951,561 user-item pairs from Nov. 2004 to Mar. 2012. This dataset is very sparse, for some users tagged a litter papers and some items are tagged only once. We got a denser dataset for experiment by removing the users who tagged less than 10 research papers and removing the items which have been tagged by less than 5 users. We select data randomly from the above dataset to form the final dataset. The final dataset contains 9,170 users, 11,343 papers, 194,596 user-item pairs.

(2) Evaluation Metric

We use Hit Ratio [195] as the top-N recommendation evaluation metric. The test method we used is All-But-One. The latest item user bookmarked or tagged is selected as the test data and other items are collected as the training data. We recommend N items to user u at time t. The set of recommended items is denoted as $R(u,t)$. T_u is the target behavior of user u in the test dataset. I is a indicate function to indicate whether the test item appears in the recommendation list. If this kind of appearance happened, we call it a hit. The formula of Hit Ratio is as follow:

$$Hit\,Ratio = \frac{\sum_u I(T_u \in R(u,t))}{|N|} \tag{3.39}$$

(3) Compared Methods

The most famous algorithms to do top-N recommendation are user-based collaborative filtering (CF) method called User-KNN [2] and item-based CF method called Item-KNN [110]. User-KNN employs statistical techniques to find a

Table 3.2 Pseudo code of DARGDARG

Input: NodeSet V, user *u*, time *t*
Output: Top-N recommendation for user *u* at time *t*

```
Queue Q;
Q.push(u);
//initialization
for each item node i in V
  distance[u, i]=OXFFFF;
  weight[u, i] = 0;
end for
While Q is not empty
   Node top = Q.top();// The first one in the queue
   Q.pop();
   for each Node v' in V
     Calculate w_d , w_op , w_cp , and w_uti
        newDistance[u, v'] = distance[top] + 1;
        newWeight[u, v']=( w_d + w_op + w_cp + w_uti )/4;
        if newDistance[u, v'] > 3
           break;
        end if
        if newDistance[u, v'] < distance[u, top]
           distance[u, top] = newDistance[u, v'];
           Q.push(top);
        else if newDistance[u, v'] = distance[u, top]
           weight[u, top] += newWeight[u, v'];
           Q.push(top);
        end if
   end for
end while
weight.sort();
return top-N unknown items;
```

collection of users, known as *neighbors*, which have a history of agreeing with the target users. Once a neighborhood of users is formed, algorithms to combine the preferences of neighbors are used to product top-N recommendations for the active user. Item-KNN techniques first analyze the user-item matrix to identify relationships between different items, and then user these relationships to indirectly produce recommendations for users.

In recent years there have been some research on temporal top-N recommendation on binary data, but got only a little attention. References [196, 120] both proposed a time weighted item-based CF method (TItem-KNN), which reduced the influence of older data when predicting user's further behaviors. The basic idea of these approaches is the user's further choices are determined by his/her recent behavior. Similarly, User-KNN method can be transformed to TUser-KNN supporting temporal top-N recommendations. The basic idea of TUser-KNN is the user's further behavior can be predicted by his/her neighbors' usage history. So we use four methods as compared algorithms, including User-KNN, Item-KNN, TUser-KNN, and TItem-KNN (Table 3.3).

(4) Parameter Setting

We try some different values of φ to find the most optimal one. As the Fig. 3.4 shows, the highest Hit Ratio is 12.814 % when φ value is 3.

Considering learner's cognitive level, we try different value of φ_1, φ_2, and σ based on the value of φ. Figure 3.5 describe that the Hit Ratio varies by different value of φ_1, φ_2, and σ. The best performance is 12.977 % when the values of φ_1, φ_2, and σ are 2, 3 and 25 correspondingly. The improvement is 1.28 % over the fixed φ.

We have not tried all the possible values exhaustively, since it may be lead to over fitting problem. Additionally, different optimization methods should be

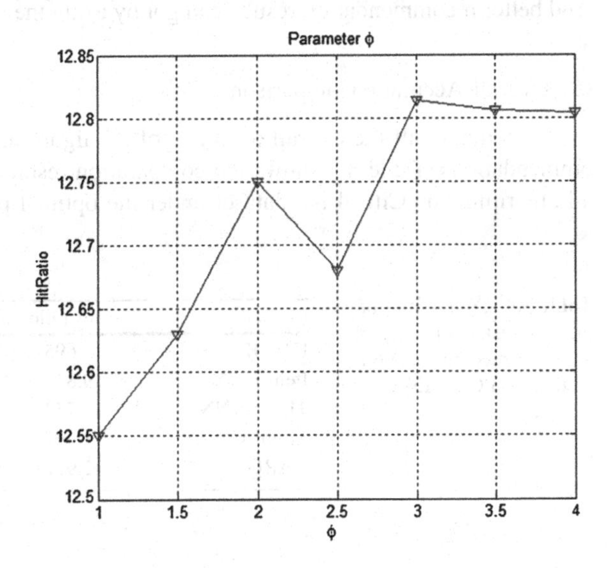

Fig. 3.4 Performance with different φ on CiteULike

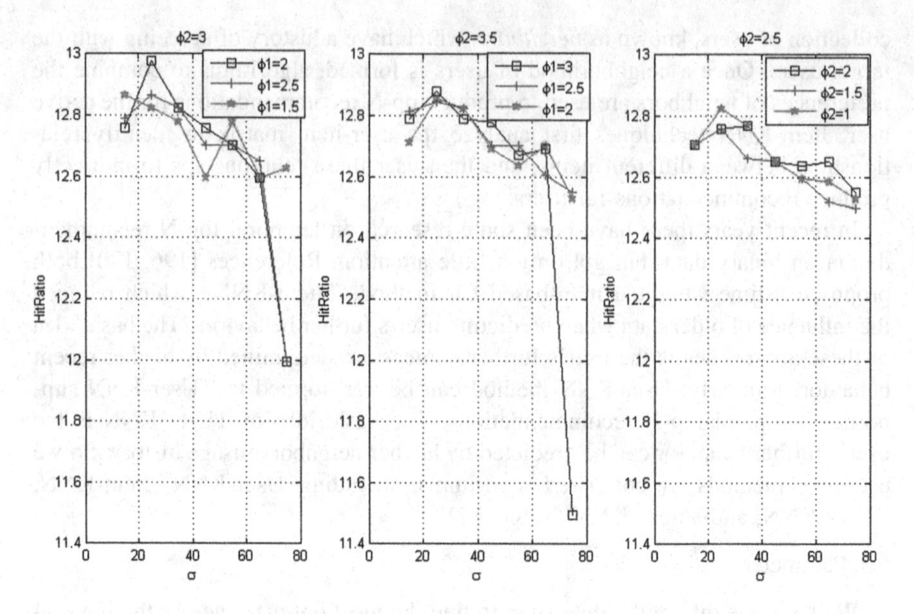

Fig. 3.5 Performance with different φ_1, φ_2, and σ on CiteULike

applied based on different applications. The aim of our study is to prove that academic recommendation accuracy can be improved by taking cognitive level into account. However, measuring learner's cognitive level based on the time span he/she using the system is somewhat rough, it can bring a certain degree of improvement of the top-N recommendation accuracy. If the practical applications can collect more user information, such as age, education level, the articles' level he/she read or tagged, the model of user's dynamic cognitive level can be built. And better recommendation results can got by using the algorithm proposed in this study.

(5) Overall Accuracy Comparison

We compare all the overall accuracy of all algorithms in temporal top-N recommendations. Table 3.3 shows the comparation results of the best Hit Ratios of all algorithms on CiteULike dataset under the optimal parameters.

Table 3.3 Comparison of the best results of different temporal recommendation methods on CiteULike dataset

Method	Hit ratio (%)	Improvement (%)
User-KNN	9.695	33.853
Item-KNN	9.836	31.934
TUser-KNN	10.752	20.694
TItem-KNN	11.497	12.873
DARG	12.977	–

Results show that User-KNN has the lowest Hit Ratio, and the performance of Item-KNN method are closer to the User-KNN. TItem-KNN and TUser-KNN got better results by taking temporal element into account. Moreover, DARG outperforms all the other compared methods. This proves that the time factors play an important role in improving recommendation accuracy. Comparing DARG with the TItem-KNN which is the best perform algorithm in the compared algorithms, DARG achieves 12.873% improvement. This proves that our DARG algorithm is effective in leveraging temporal dynamics for top-N academic recommendation based on implicit data.

3.6 Conclusion

In this case study, we proposed a model named Dynamic Transfer Chain (DTC) to capture the user's preferences and academic trend over time. Based on DTC framework, we present a novel algorithm Dynamic Academic Recommendation on Graph (DARG) to do top-N academic recommendation on implicit data. The experiments on real datasets confirm the effectiveness of the proposed algorithms over other existing methods. Although the DARG is designed for academic recommendation, other applications can also benefit from it, for it takes four time factors about users and items into account. Future work includes using more accurate cognitive approach to model learner's cognitive level, and then proves whether the recommendation accuracy can be improved by using it. In addition, we also plan to explore top-N academic recommendation on explicit data using the proposed DTC.

Chapter 4
Sentiment Analysis

Sentiment analysis (also called opinion mining) refers to the application of natural language processing, computational linguistics, and text analytics to identify and classify subjective opinions in source materials (e.g., a document or a sentence). Generally speaking, sentiment analysis aims to determine the attitude of a writer with respect to some topic or the overall contextual polarity of a document. The attitude may be his or her judgment or evaluation, affective state (that is to say, the emotional state of the author when writing), or the intended emotional communication (that is to say, the emotional effect the author wishes to have on the reader).

Generally, sentiment analysis classifies text expressions in source materials into two types: (1) facts (objective): objective expressions about entities, events and their attributes, e.g., "I bought an iPhone yesterday"; and (2) opinions (subjective): subjective expressions of sentiments, attitudes, emotions, appraisals or feelings toward entities, events and their attributes, e.g., "I really love this new camera". It should be pointed out that not all subjective sentences contain opinions, e.g., "I want a phone with good voice quality"; and not all objective sentences contain no opinions, e.g., "The earphone broke in just two days!"

Therefore, it is important for sentiment analysis to identify and extract facts and opinions from source text materials. However, unfortunately, this is difficult to be achieved accurately. Let us consider the following example, which is a simple review on iPhone.

Example 1.1 (1) I bought an iPhone a few days ago. (2) It was such a nice phone. (3) The touch screen was really cool. (4) The voice quality was clear too. (5) Although the battery life was not long, that is ok for me. (6) However, my mother was mad with me as I did not tell her before I bought it. (7) She also thought the phone was too expensive, and wanted me to return it to the shop □.

In the above example, it is clear that the sentence (1) would be identified as a fact; and the sentences (2), (3) and (4) as opinions. However, for the sentences (5), (6) and (7), it may be difficult to automatically determine the opinion expressions on iPhone. Generally, opinions have targets (objects and their attributes) on which

T. Luo et al., *Trust-Based Collective View Prediction*,
DOI: 10.1007/978-1-4614-7202-5_4, © Springer Science+Business Media New York 2013

opinions are expressed. Below, we give some definitions to formally describe opinions [125].

Definition 1.1 (*Object*) an entity that can be a product, service, individual, organization, event, or topic, e.g., iPhone.

Definition 1.2 (*Attribute*) an object usually has two types of attributes: (1) components, e.g., battery, keypad/touch screen; and (2) properties, e.g., size, weight, color, voice quality.

Definition 1.3 (*Explicit and implicit attributes*) explicit attributes refer to those appearing in the opinion, e.g., "the battery life of this phone is too short"; and implicit attributes refer to those not appearing in the opinion, e.g., "this phone is too large" (on attribute size).

Definition 1.4 (*Opinion holder*) the person or organization that expresses the opinion.

Definition 1.5 (*Opinion orientation*) i.e., polarity, e.g., positive, negative or neutral.

Definition 1.6 (*Opinion strength*) level/scale/intensity of opinion indicating how strong it is, e.g., contented, happy, joyous and ecstatic, whose strength are incremental.

Definition 1.7 (*Opinion*) a person or organization that expresses a positive or negative sentiment on a particular attribute of an object at a certain time, thus, which can be represented as a quintuple: < object, attribute, orientation, opinion holder, time > .

However, in Definition 1.7 about the quintuple on opinions, it should be pointed out that (1) some information may be implied due to pronouns, context, or language conventions; (2) some information available from document attributes; and (3) in practice, not all five elements are needed. Besides, options can be classified into: (1) direct opinion, sentiment expressions on one or more attributes of an object, e.g., "the voice quality of this phone is fantastic"; and (2) comparative opinion, relations expressing similarities or differences between two or more objects based on some of the shared attributes of the objects, e.g., "the voice quality of camera x is better than that of camera y". However, from the viewpoints of opinion expressions, they also can be can be classified into: (1) explicit opinion, an opinion on an attribute explicitly expressed in a subjective sentence, e.g., "the voice quality of this phone is amazing"; and (2) implicit opinion, an opinion on an attribute implied in an objective sentence, e.g., "the headset broke in two days".

A main goal in sentiment analysis is to identify and classify options in source text expressions, i.e., to identify whether given document (e.g., product reviews, blogs, forum posts) or a given sentence expresses opinions and whether the opinions are positive, negative, or neutral. More generally, the basic tasks of sentiment analysis mainly includes: (1) sentiment identification, i.e., subjectivity identification, which aims to identify whether a piece of text expresses opinions;

and (2) sentiment orientation classification, which aims to determine the orientation of an opinionated text. In the following sections, we briefly review some methods on the above two tasks.

4.1 Sentiment Identification

Sentiment identification (also called subjectivity identification) refers to identify whether a piece of text (e.g., a document, a sentence) expresses opinions, generally, which is based on the following two basic assumptions: (1) the given text is opinionated on a single object; and (2) the opinions are from a single opinion holder.

In sentiment identification, the main task is to identify opinion words, which is very important. Opinion words are dominating indicators of sentiments, especially adjectives, adverbs, and verbs, e.g., "I absolutely love this camera. It is amazing!" Opinion words are also known as polarity words, sentiment words, opinion lexicon, or opinion-bearing words, which generally can be partitioned into two types: (1) positive words, e.g., wonderful, elegant, amazing; and (2) negative words, e.g., horrible, disgusting, poor. In order to identify opinion words from a given piece of text, we need to in advance generate a set of opinion words. If we have known a rich set of opinion words, the sentiment identification process is easy to be achieved, which only need to search out each opinion word from source text expressions.

Aiming at generating opinion words, there are three types of methods: (1) manual generation method, i.e., collecting opinion words manually, obviously, which is effective (i.e., accurate) but expensive; (2) dictionary-based generation method, which uses a seed list and grow the list, e.g., SentiWordNet [126]; and (3) corpus-based generation method, i.e., relying on syntactic or co-occurrence patterns in large text corpora. Below, we briefly review the latter two types of methods (i.e., dictionary-based and corpus-based) on how to generate opinion words.

4.1.1 Dictionary-Based Opinion Words Generation

In this method, a list of seed opinion words is needed to prepare in advance; and then a dictionary is used to help to grow the list to generate more opinion words. More specifically, the method of dictionary-based opinion words generation can be described as follows:

1. A small seed set of opinion words with known orientations is collected manually, e.g., {"glad"}, a set of a positive opinion word "glad".

2. An online dictionary (e.g., WordNet [127]) is then searched for their synonyms and antonyms to grow the seed set of opinion words, for example, after adding the synonyms of "glad", we obtain a set of more positive opinion words: {"glad", "happy", "joyful", "delighted"}; and after adding antonyms of "glad", we obtain a new set of negative opinion words: {"sad", "unhappy", "sorry", "heart-broken"}.
3. The above two steps would be repeated until no more new opinion words can be found from the online reference dictionary.
4. Finally, manual inspection may be done for correction.

In addition, in the above process, some additional information (e.g., glosses) from WordNet can be used. Overall, the process of dictionary-based opinion words generation is simple and easy to be understood. However, this method cannot identify context-dependent opinion words, i.e., the words whose opinion orientations heavily rely on the context. Let us consider the following two examples.

Example 2.1 Given an opinion word "small", and two sentences "The LCD screen is too small" and "The camera is very small and easy to carry", we can find that in the first sentence, the opinion word "small" is negative, and in the second sentence, it is positive, i.e., the opinion orientation of the word "small" is context-dependent □.

Example 2.2 Given an opinion word "long", and two sentences "It takes a long time to focus" and "The battery life is long", we can find a situation similar to Example 2.1 is encountered □.

4.1.2 Corpus-Based Opinion Words Generation

Different to the dictionary-based generation, corpus-based opinion words generation depend on syntactic or co-occurrence patterns in large text corpora [128]. One of the main advantages of corpus-based opinion words generation over dictionary-based generation is that it can obtain domain dependent orientations and/or context dependent ones [125]. Below, we give an example of identifying the sentiment orientations of adjectives in context.

1. Start with a list of seed opinion adjective words.
2. Based on a chosen corpus, use linguistic constraints on connectives to identify additional adjective opinion words and their orientations, where the linguistic constraints include: (1) sentiment consistency, which is based on the observation that conjoined adjectives usually have the same orientations, e.g., given a sentence "This car is beautiful and spacious", if the word "beautiful" is positive, then "spacious" is positive too. (2) some rules can be designed for different connectives, e.g., AND, OR, BUT, EITHER-OR, NEITHER-NOR.
3. Use log-linear model to determine if two conjoined adjectives are of the same or different orientations.

4. Use clustering to produce two sets of opinion words, i.e., positive words and negative words.

Using the above process, we can determine domain opinion words. However, finding domain opinion words is not sufficient. One opinion word may indicate different opinions in the same domain, e.g., "The battery life is long" versus "It takes a long time to focus". To overcome this problem, a basic idea is to identify all opinion words related to a specific object attribute, whose process can be briefly described as follows (see [129] for more detail): (1) create pairs of object attribute and opinion word, i.e., < object attribute, opinion word > ; and (2) then, determine opinion words and their orientations together with the object attributes. Based on the above process, the context dependency of opinion words can be handled.

4.2 Sentiment Orientation Classification

Sentiment orientation classification refers to determine the opinion orientation of an opinionated text, i.e., based on the opinion words identified from a given piece of text by sentiment identification, to determine whether the opinion orientation in the given text are positive, negative or neutral.

Based on the assumption that the given text is opinionated on a single object, that is to say, all the opinion words identified from the given text act on a single object, a straightforward way of classifying sentiment orientation is to count positive and negative opinion words, e.g., in Example 1.1, the number of positive opinion words on the object "iPhone" is greater than the number of negative words, so it can be simply considered that the review expresses positive sentiment orientation. Besides, machine learning approaches also can be used for sentiment orientation classification. Below, we briefly review two types of methods on sentiment orientation classification (i.e., counting opinion words and supervised machine learning).

4.2.1 Counting Opinion Words

Counting opinion words is a simple method on sentiment orientation classification, which is based on the predefined opinion words generated by sentiment identification. In this method, first, we need to assign orientation score (+1, −1) to all opinion words: (1) positive opinion words (+1), e.g., great, amazing, love; and (2) negative opinion words (−1), e.g., horrible, hate. In this process, we also can use strength value between [0, 1], based on the opinion strength of opinion words. Last, the opinion orientation score of a given piece of text is simply considered to be equal to the sum of orientation scores of all opinion words found. Let us consider the review on iPhone in Example 1.1. If we assign (+1) to each positive opinion word (i.e., "nice", "cool", "clear" and "ok") and (−1) to the negative opinion word(i.e., "expensive"), then the review has an opinion orientation score of $4-1=3$, and thus it has positive opinion orientation.

However, the above way by simply counting opinion words is obviously not accurate enough, for example, given a sentence "There is not one thing I hate about this product", it would be assigned with a negative opinion orientation score, but in fact it is positive. To overcome the problem caused by negation words, we can create some basic opinion rules to improve the accuracy of simply counting opinion words, for example, two simple rules can be manually created as follows: both "not ... negative" and "never ... negative" represent positive opinions. See [125] for more detail on rule-based opinion words counting methods.

4.2.2 Supervised Learning Approaches

Using counting opinion words, generally, only a limited number of opinion words can be found, and only a limited number of patterns can be created. Can we automate the task with limited manual work (e.g., find opinion words and their orientations automatically)? To solve this, supervised learning approaches are advocated to apply into sentiment orientation classification. Here, the basic idea is to leverage supervised learning techniques to find patterns in known examples and apply them to new documents, so as to classify the sentiment orientation of new documents automatically. Thus, the goal of supervised learning is to train and obtain an opinion classifier which contains some target opinion classes, e.g., positive versus negative. Most existing supervised learning approaches can be used to achieve the goal (i.e., to obtain the opinion classifier). Now, popular supervised learning methods mainly include:

1. Naïve Bayes (NB): A simple probabilistic classifier based on applying Bayes' theorem with strong (naive) independence assumptions.
2. Maximum Entropy (ME): A probabilistic model that estimates the conditional distribution of the class label.
3. Support Vector Machines (SVM) [130]: A representation of the examples as points in space in which support vectors are computed to provide a best division of points/examples into categories.
4. Logistic Regression Model (LR) [131]: A LR model predicts the classes from a set of variables that may be continuous, discrete or a mixture.

However, a classifier obtained by supervised learning approaches often lead to a problem of domain dependency, i.e., the classifier trained using opinionated documents from domain A often performs poorly when tested on documents from domain B. Generally, this problem is caused due to the following two reasons [132]: (1) words used in different domains can be substantially different, e.g., Cars versus movies, and Cameras versus Strollers; and (2) some words mean opposite in two domains, e.g., "unpredictable" may be negative in a car review, but positive in a movie review, and "cheap" may be positive in a travel/lodging review, but negative in a toys review. This problem can be solved by the approach in [133], which is briefly described as follows :

1. Use labeled data from one domain and unlabeled data from both source the target domain and general opinion words as features.
2. Choose a set of pivot features which occur frequently in both domains.
3. Model correlations between the pivot features and all other features by training linear pivot predictors to predict occurrences of each pivot in the unlabeled data from both domains.

4.3 Case Study: Sentimental Analysis in Recommender Systems

4.3.1 Introduction

Recently, social tagging systems originated with sites such as Del.icio.us[1] and Flickr[2] have become a popular trend in Web 2.0 environment. Social tagging systems are designed for the storage, organization, retrieval and share of personal resources such as links, videos and photos on the web. By tagging, users annotate and index the interested resources freely and subjectively, based on their senses of interests. As such, these user generated tags can be seen as a new metadata carrying on the user preference and resource relatedness in social collaborative systems.

Nowadays people are inundated by information loads and choices. Recommender systems are proposed as a solution to this problem by providing users with the interested and needed information. Various kinds of recommender systems [134, 135] are developed for better user experience, in which the personalized recommendation takes an important aspect within the whole recommendation approaches. Traditional recommender systems focus on the explicit rating data of users, e.g., movie ratings, to gain the user-interest profile and make predictions for new items. Different from the rating data, social tagging data does not contain user explicit preference information on resources, instead, reflecting the implicit perceptions on certain resources by annotating their opinions or perceptions. Therefore, making use of tagging data as an additional feature will undoubtedly facilitate the capture of user preference and resource relatedness for better personalized recommendation. Recently social tag recommender system is emerging as an active research topic in the domain of recommender systems.

---

[1] http://del.icio.us
[2] http://flickr.com

As discussed above, the aim of recommendation is to improve the user experience and satisfaction when they are browsing the web and interacting with various systems for specific information needs. Thus in recommendation practice we sometimes need to recommend the items or resources that gained positive feedbacks or reflected preferable comments in addition to the intuitive similarities on some aspects such as categorical and contextual information. The tagging data that annotated by users on resources, of course, is able to provide such kind of polarity attributes in social annotation systems. Intuitively, we aim to incorporate sentiment analysis into social tag recommender systems to improve current recommendation approaches.

Some related studies have been conducted to reveal the subjective inherence of tagging in terms of sentiment (or opinion) expression. For example, Golder and Huberman [136] found evidences of sentiment expression in the self-tagging site Del.icio.us. Zollers [137] discovered the sentiment expression in social tagging systems like Amazon.com and Last.fm (shown in Tables 4.1 and 4.2).

All of the sentimental tags found in Last.fm and Amazon.com along with other co-occurred noun tags which described the resource that was being tagged indicate the polarized opinions of users on these resources. Additionally, it is possible to see that the expressed opinions also work as indicators in recommendations either for or against a resource.

Traditionally, sentiment analysis refers to classifying text documents, such as user reviews, newsgroup messages and blogs, based on the polarity of the opinions they express [47]. Specifically, sentiment analysis is concerned with the automatic identification, extraction, and classification of opinions in texts. It can be used to develop applications that assist decision makers and information analysts in tracking user opinions about topics that they are interested in. Examples of sentiment analysis include the classification of a movie review as "thumbs up" or "thumbs down" [132].

Motivated by the intuition of sentiment analysis, in this research we aim to propose a new recommendation approach by incorporating sentiment factor of tagging into social tag recommender systems. On top of conventional social tag recommender systems, the tag sentiment information presented in the form of

Table 4.1 Sample opinion tags found on last.fm

Opinion tag	Tag data
Awesome	7, 034 people used this tag 36, 363 times
Beautiful	6, 865 people used this tag 38, 510 times
Great	2, 028 people used this tag 7, 756 times
Crap	1, 520 people used this tag 4, 950 times

Table 4.2 Sample opinion tags found on Amazon.com

Opinion Tag	Tag data
Awesome	691 people used this tag on 855 items
Beautiful	299 people used this tag on 356 items
Cool	404 people used this tag on 574 items

subjective polarity of the user toward the annotated resources, work like an additional information filtering to recommend more preferable and positive resources to the user. Therefore, the sentimental tags can be used to improve the performance of recommendation. To our best knowledge, there is very little work addressing the sentiment analysis enhancement for improved tag-based recommendations in previous studies.

4.3.2 Related Work

4.3.2.1 Tag-Based Personalized Recommendation

Durao and Dolog [138] developed a multi-factorial tag-based recommender system, which took various lexical and social factors of tags into the similarity calculation. Using tags as a means to express which features of an item user particularly like or dislike, Gedikli and Jannach [139] proposed a simple recommendation method that can take item-specific tag ratings into account when generating rating predictions. Zhao et al. [140] proposed a collaborative filtering approach Tag-based Collaborative Filtering (TBCF) based on the semantic distance among tags assigned by different users to improve the effectiveness of neighbor selection. Shepitsen et al. [141] proposed a personalized recommendation system by using hierarchical clustering. In this approach, instead of using the pure tag vector expressions, a processing on tag clustering was performed to find out the tag aggregates for personalized recommendation. Xu et al. [142] proposed a Semantic Enhancement Recommendation strategy (SemRec), based on both structural information and semantic information through a unified fusion model, which combines the clustering and hidden topic model. Extensive experiments conducted on two real datasets demonstrate the effectiveness of their approaches. Leung et al. [143] described a rating inference approach to incorporating textual user reviews into collaborative filtering (CF) algorithms. The main ideas of their approach is to elicit user preferences expressed in textual reviews, a problem known as sentiment analysis, and map such preferences onto some rating scales that can be understood by existing CF algorithms.

4.3.2.2 Sentiment Analysis

In the literature, sentiment analysis goes under various names, such as opinion mining, sentiment mining. Its related work may come from both computer science and linguistics. The task of sentiment analysis can be roughly divided into three sub-categories: determining subjectivity [144], determining orientation, and determining strength of orientation [145], and most of the studies focus on investigating the sentiment orientation of words, phrases, and documents. Turney [132] used point-wise mutual information (PMI) to calculate an average semantic orientation score of extracted phrases for determining the document's polarity. Amps et al. [146] tried to evaluate the semantic distance from a word to good/bad with WordNet. Pang et al. [130] employed three machine learning approaches to annotate the polarity of IMDB movie reviews. Jindal and Liu [48] built a framework to compare consumer opinions of competing products using multiple feature dimensions. After deducting supervised rules from product reviews, the strength and weakness of the product were visualized with an Opinion Observer. Scaffidi et al. [145] presented a search system called Red Opal that examined prior customer reviews, identified product features, and then scored each product on every feature. Zhuang et al. [147] presented a supervised approach for extracting feature-opinion pairs. Their method learned the opinion and a combination of dependency and part-of-speech paths connecting such pairs from an annotated dataset.

4.3.3 Preliminaries

4.3.3.1 Social Tagging System Model

In this study, our work is to deal with tagging data. A typical social tagging system has three types of entities, users, tags and resources which are interrelated with one another. Social tagging data can be viewed as a set of triples [148]. Each triple (U_i, T_j, R_k) represents an observation of a user U_i annotating a tag T_j on a resource R_k. A social tagging system can be described as a four-tuple—there exist a set of users, U; a set of tags, T;a set of resources, R; and a set of annotations, A. We denote the data in the social tagging system as D and define it as: $D = \langle U, T, R, A \rangle$. The annotations, A, are represented as a set of triples containing a user, tag and resource defined as: $A \subseteq \langle U_i, R_j, R_k \rangle : U_i \in U, T_j \in T, R_k \in R$.

4.3.3.2 Standard Tag-Based Recommendation

The standard tag-based recommendation is principally similar to a process of traditional information retrieval but with an additional input of the user tagging preference for personalization (or called personalized recommendation). The procedure consists of two steps of search and personalization. The search step produces a list of candidate resources based on the similarity computation between the query tag issued by a user and all resources in terms of term frequency—inverse document frequency (tf-idf).

The second step utilizes the tagging preference of users to make the personalization. Under the vector space model, each user, u, is modeled as a vector (also called user profile) over the set of tags, where $w_i(T_i)$, in each dimension corresponds to the relationship of a tag t_i with this user, U_i $U_i = \langle w_i(T_1), w_i(T_2), \cdots, w_i(T_{|T|}) \rangle$.

Likewise each resource, R_j, can be modeled as a vector (i.e., resource profile) over the same set of tags, $R_j = \langle v_j(T_1), v_j(T_2), \cdots, v_j(T_{|T|}) \rangle$. After that, the similarity computation, e.g., cosine measure, of the target user profile u and the candidate resource profiles r selected by the first step, is performed, $sim(U_i, R_j)$, to further generate the personalized resources based on various recommendation strategies. The distinction of the tag-based recommendation from the standard information search is that here the recommendation is derived upon, not only the query itself, but also the user tagging preference (i.e., personalization).

4.3.4 Sentiment Enhanced Approach for Tag-Based Recommendation

4.3.4.1 Sentiment Enhanced Approach

In order to generate personalized recommendations, Durao et al. proposed a framework for the calculation of similarity between resources based on tags. Their method combines the basic cosine similarity calculus with other factors, such as tag popularity, tag representativeness and an affinity user-tag for the purpose of reordering the original raking in recommendation and generates personalized ones. Based on their work and incorporating a sentiment factor, we present an approach to calculating the similarity of resources as follows:

$$Sim(R_A, R_B) = [(D_A + D_B) * cos\_sim(R_A, R_B)] * sentiment(UR_A, \ \mathrm{R}_B) \qquad (4.1)$$

where, R_A and R_B are the resources in a social tag system. If a user *UR* labels the resource R_A by tagging, and resource R_B is the other resource in the system, Eq. (4.1) calculates the similarity score between R_A and R_B. We then select the Top-N score resources for the user *UR* who labels the resource *A* as the personalized recommendation.

In Eq. (4.1), D_A and D_B is the score of resource R_A and R_B, D_A is defined as follows:

$$DA = \sum_{i=1}^{n} (weight(Ti) * representativness(Ti)) \qquad (4.2)$$

where n is the total number of tags labeled on R_A. The $weight(T_i)$ factor is the popularity of the T_i in the social tagging system, which is calculated as a count of occurrences of one tag per total of resources available. We rely on the fact that the most popular tags are like anchors to the most confident resources. As a consequence, it decreases the chance of dissatisfaction by the receivers of the recommendations.

The $representativeness(T_i)$ factor measures how much a tag can represent a document it belongs. It is believed that those tags which appear most in the document can better represent it. The tag representativeness is measured by the term frequency, a broad metric also used by the information retrieval community.

The $\cos\_sim(R_A, R_B)$ factor in Eq. (4.1) is a cosine similarity between resource R_A and R_B from the classical text mining and information retrieval domain, where, two resources are thought of as two vectors in the m-dimensional tag-space under the vector space model. The similarity between them is measured by computing the cosine function of the angle between these two vectors.

The $sentiment(UR_A, R_B)$ in Eq. (4.1) is a sentiment enhanced factor proposed in our approach, which is defined as follows:

$$sentiment(URA, R_B) = \sum_{i=1}^{n} sentiment\_score(Ui, R_B) \tag{4.3}$$

where, $UR_A = \{U_1, U_2, \ldots, U_N\}$ is a set of Top-N nearest neighboring users of user UR who labels the resource R_A. The similarity between two users is calculated using cosine similarity based on tag-vector space model same as $\cos_sim(R_A, R_B)$factor.

$sentiment\_score(U_i, R_B)$ in Eq. (4.3) is the total sentiment score of user U_i towards resource R_B as follows:

$$sentiment\_score(Ui, R_B) = \sum_{i=1}^{k} sentiwordnet(Ti) \tag{4.4}$$

where, T_i is a tag labeled on resource R_B by user $U_i \cdot sentiwordnet(T_i)$ is a polarity score of T_i calculated based on the Senti WordNet, a public available resource for sentiment analysis evolved from Wordnet, which is a wildly used lexical database of English. In SentiWordNet, each WordNetsynset is assigned with a triplet of numerical scores representing how Positive, Negative and Objective a synset is. SentiWordNetis proved as a useful tool for sentiment mining applications, because of its wide coverage (all WordNetsynsets are assigned according to each of the three labels Objective, Positive, Negative) and because of its fine granularity, obtained by qualifying the labels by means of numerical scores.

In our approach, $sentiwordnet(Tag_i)$ in Eq. (4.4) is defined as follows:

$$sentiwordnet(T_i) = \max_{positive}(T_i) - \max_{negative}(T_i) \tag{4.5}$$

where, $\max_{positive}(T_i)$ and $\max_{negative}(T_i)$ are the maximum positive and negative polarity score of synsets of T_i Normally, a tag have a set of different synsets in Wordnet, therefore, T_i have multiple different scores of positive and negative polarity score for each synset. In order to emphasize the effect of the sentiment enhanced factor of a tag labeled on the resource, we choose the maximum positive and negative polarity score of synsets of T_i as calculating parameters in Eq. (4.4).

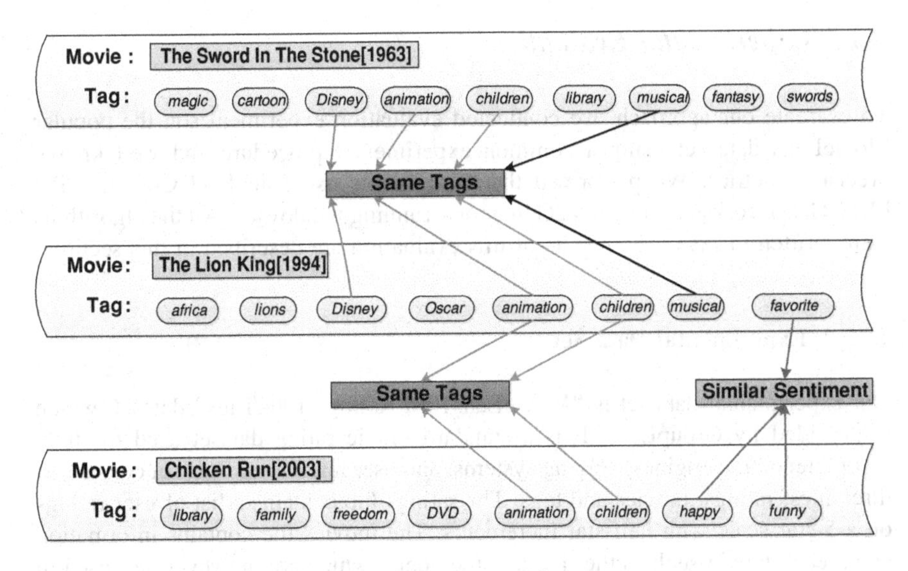

Fig. 4.1 Example of sentiment enhanced tag recommendation

4.3.4.2 Running Example

Now, let us look at a running example which explains our approach and ideas behind it. According to Fig. 4.1, three movies "The Sword In The Stone [1963]", "Chicken Run [2003]", "The Lion King [1993]" and their tags taken from MovieLens are shown. In this example, The "The Lion King" and "Sword In The Stone" have four same tags. "The Lion King" and "Chicken Run" only have two same tags, but they own three sentiment tags with same positive polarity. The polarity scores of three tag calculated from Eq. (4.5) are shown in Table 4.3. Here, one synset of the tag "*funny*" is "experiencing odd bodily sensations", therefore, the $\max_{negative}(funny)$ value of the word "funny" is larger than $\max_{positive}(funny)$ value.

If the sentiment enhanced factor is not considered, *Similarity* [The Lion King (1993), The Sword In The Stone (1963)] is larger than *Similarity* [The Lion King (1993), Chicken Run (2003)], otherwise, *Similarity* [The Lion King (1993), Chicken Run (2003)] is larger than *Similarity* [The Lion King (1993), The Sword In The Stone (1963)]. Therefore, Chicken Run (2003) will be recommended to the user who labels The Lion King (1993) in our approach, rather than The Sword In The Stone (1963).

Table 4.3 Sentiment polarity scores of tags

Tag	Positive score	Negative score
Favorite	0.375	0.125
Happy	0.875	0
Funny	0.5	0.625

4.3.5 Experimental Evaluation

To evaluate our approach, we conducted evaluation experiment son the popular MovieLens data set using a common experimental procedure and well known precision metrics. We performed the experiments using an Intel Core i3 CPU (2.13 GHz) workstation with a 2G memory, running windows 7. All the algorithms were written in Java. The results of this evaluation are described in this section.

4.3.5.1 Experimental Data Set

Our experimental data set is "MovieLens10M Ratings 100kTags" dataset, which is provided by GroupLens. It is a standard movie rating dataset used to study recommendation engines, tagging systems, and user interfaces. Data set consists of three files: ratings, movies and tags. The ratings file contains a list of user ratings on a 5-star scale with half-star increments. The movies file contains information about each movie such as the title and the genre, which are, however, not used in our method. The tags file contains the information about which tags have been assigned by the users to the movies. A tag assignment is a triple consisting of one user, one movie an done tag. Data set contains 10, 000, 054 ratings and 95, 580 tags applied to 10, 681 movies by 71, 567 users of the online movie recommender service MovieLens.

4.3.5.2 Data Set Preprocessing

Tag quality is one of the major issues when developing and evaluating approaches that operate on the basis of user-contributed tags. Reference [149] revealed that only 21 % of the tags in the MovieLens system had adequate quality to be used to evaluate approach.

A data pruning measures is applied to improve the quality of the existing tag information in our experiment. We defined the following two requirements for movies, tags and users to be taken into account in our evaluation. First, we only consider movies that have at least 15 tags assigned. Second, only those users are considered that they rate at least 10 movies to avoid problems with memory limitations. The resulting dataset used in our experiments finally includes 3,390 movies, 1,151 users and 2,645 tags shown in Table 4.4.

Table 4.4 Statistics of experiment datasets

Number of users	1,151
Number of movies	3,390
Number of tags	2,645
Minimum movies per user	10
Minimum tags per movies	15

4.3.5.3 Evaluation Methodology and Results

The goal of our analysis is to determine whether the sentiment enhanced recommending approach is effective. We compared the following algorithms:

- TBR-CS(tag-based recommender with cosine similarity): it uses the similarity metric $Sim(R_A, R_B) = [(D_A + D_B) * cos\_similarity(R_A, R_B)]$ in Eq. (4.1).
- TBR-AF(tag-based recommender with affinity factor): it uses the similarity metric $Sim(R_A, R_B) = [(D_A + D_B) * cos\_similarity(R_A, R_B)] * AF(U_i, T_j)$ as Durao et al. proposed.
- TBR-SF(tag-based recommender with sentiment factor): it uses the similarity metric
- $Sim(R_A, R_B) = [(D_A + D_B) * cos\_sim(R_A, R_B)] * sentiment(UR_A, R_B)$ as Eq. (4.1) proposed.

We utilized the standard metric from the area of information retrieval to evaluate our approaches. For each dataset, we randomly and judiciously divided the whole dataset into two parts by 80 (Training set) and 20 % (Test set). Here we use precision as evaluation metrics. In precision evaluation, for each given user from the test set, we determine the Top-N resources as recommendations based on the generated sentiment enhanced similarity scores. Then we count the total number of resources which are simultaneously occurred in the recommended resource list and real test set for each user and calculate the ratio of this number to the recommendation size as precision. Eventually we average the precision values over the total test set to obtain the final evaluation result.

Here, we report the experimental results of improved recommendation performance in comparison to two baseline approaches, i.e., using pure cosine similarity metric(TBR-CS) and affinity enhanced metric(TBR-AF). With the parameter N = 50, an example with the top 5 movies recommended to the user identified by User-ID 146 who annotated the movie "Star Wars Episode I" by three algorithms is shown in Table 4.5. From the example, we can see that the recommendation made by approach TBR-AF looks more reasonable, e.g., "*Star Wars Episode IV*" is ranked at the top.

The overall precision results of three recommendation approaches are shown in Table 4.6. From Table 4.6, we can see that our (TBR-SF) approach proposed in this study consistently outperforms both pure cosine similarity based method as

Table 4.5 Top 5 recommendation results to User-ID 146

Approach	REC	TBR-CS	TBR-AF	TBR-SF
User-ID 146/	@top-1	X-men	Stargate	Star wars episode IV
Star wars episode I	@top-2	Terminator	Star wars episode IV	X-men
	@top-3	Stargate	X-men	Terminator
	@top-4	Alien	Terminator	Stargate
	@top-5	Star wars episode IV	Alien	Alien

Table 4.6 Overall precision of three approaches

Approach	Precision (%)
TBR-CS	25.12
TBR-AF	27.31
TBR-SF	29.70

well as the affinity enhanced method in terms of precision, and the improvements are around 18.2 and 8.7 %, respectively. As a result, we conclude the proposed sentiment enhancement approach is able to achieve the better recommendation outcomes.

Chapter 5
Theoretical Foundations

Collective view is the result of collecting and processing the internet reviews. The specific form of the reviews can be unstructured text in BBS, blog and social network, or structured form in reviewing web site. We can use sentiment analysis technology [1, 47] to translate unstructured comments into structured ones. Therefore, we only research on structured comment in this book without losing of generality.

Formally, the review is a collection of author, item, text, rating (see Fig. 5.1). The item contains a number of basic attributes, such as address, classification, and some key attributes, such as taste, environment and service. When the registered users post a review, they should give an overall rating, ratings on key attributes and a short review text. System collects all the user reviews on a particular item and then gets an overall evaluation by a certain approach which we called collective view. Take the collective view on a restaurant in Fig. 5.1 for example: the overall evaluation of this shop is 5, which means "very good". More detailed evaluations for users to reference are displayed beside the overall evaluation. According to our former definition, the presentation of the results belongs to the category of collective view analysis. To simplify the analysis, we assume that all the key attributes are independent, and can be processed separately. Under this assumption, the method to analyze group insights on a specific attribute and the overall opinion are equivalent.

A simple analysis method is to get a unique result by averaging all the ratings. In this way, popular opinion will be dominant in the analysis results. However, there are two limitations of the popularity-based prediction method. First, there are differences between users' opinions, so it is difficult to obtain a satisfactory conclusion for all uses based on a uniform standard. Second, this approach ignores the reliability of comment itself and can be exploited by attackers easily. As mentioned above, more and more general public and businesses have recognized the role of internet reputation. So some companies or some people use fake web comments to enhance their brand image or combat competitors. From users' perspective, whether the system can filter these bad reviews and show the real insights of groups will determine the user's loyalty of the system. From the system owner's point of view,

T. Luo et al., *Trust-Based Collective View Prediction*,
DOI: 10.1007/978-1-4614-7202-5_5, © Springer Science+Business Media New York 2013

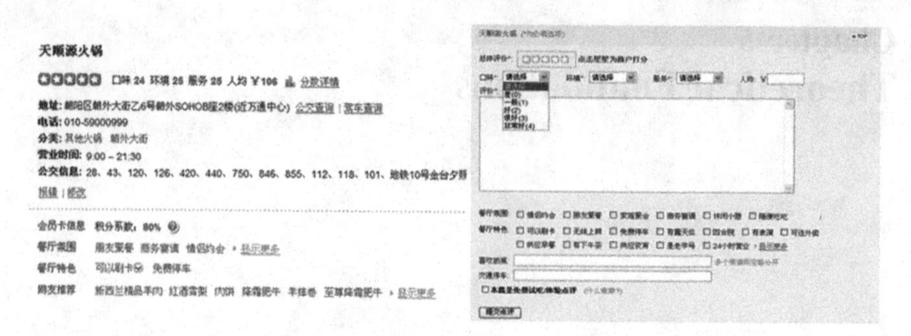

Fig. 5.1 The structure of internet comment (reference to www.dianping.com)

the overflow of false reputation information can lead to user run off, and the advertising revenue which they need for surviving will be difficult to sustain. Therefore, it is critical to find a collective view analysis method which can screen false comments and take the difference of the user's subjectivity into consideration. This method takes group reviews and target user's history in the system as input, and provides ratings of some unknown items to the target user. We call the computation process from known to unknown as collective view prediction, in order to distinguish it from non-personalized collective view analysis.

Although traditional personalized prediction method, including content-based recommendation and collaborative filtering, can capture the user's subjective differences, but they still have some certain limitations on collective view prediction problem.

5.1 Traditional Prediction Method

Content-based recommendation approach makes user of the text features of item. It selects the new comments which are similar with the comments he/she had trusted, and gives a higher weight to these comments. Unfortunately, it is difficult to distinguish false comments from the actual comments. The reason is that either false or actual reviews are about the same thing, and therefore they are similar on text characteristics, for example, both contain a large number of user's prefer terms. Of course, there may be some significant difference between fake comments and actual reviews, but the comments which once recognized by the user may not be all positive or all negative, so it is difficult to determine the reliability of comments effectively based on this method.

Collaborative filtering technology is not limited to the content itself, but some studies showed that attacker can manipulate the system output by creating fake rating record [6, 7]. We take Pearson correlation similarity as an example to explain the attack procedure. First, the similarity between user x and user z is defined as:

$$s_{xz} = \frac{\sum\limits_{\{y|\rho_{xy},\ \rho_{zy} \neq \phi\}} (\rho_{xy} - \bar{\rho}_x) \cdot (\rho_{zy} - \bar{\rho}_z)}{\sqrt{\sum\limits_{\{y|\rho_{xy},\ \rho_{zy} \neq \phi\}} (\rho_{xy} - \bar{\rho}_x)^2 \cdot \sum\limits_{\{y|\rho_{xy},\ \rho_{zy} \neq \phi\}} (\rho_{zy} - \bar{\rho}_z)^2}} \tag{5.1}$$

where ρ_x is the user x 's rating to item y. $\bar{\rho}_x$ is the average rating of the user. The computation collection is the intersection of the ratings of user x and user z. Then the rating of user x on the unknown item y is calculated by the following prediction formula:

$$\rho'_{xy} = \bar{\rho}_x + \frac{\sum\limits_{\{z|\rho_{zy} \neq \phi \wedge s_{xz} \neq 0\}} (\rho_{zy} - \bar{\rho}_z) \cdot s_{xz}}{\sum\limits_{\{z|\rho_{zy} \neq \phi \wedge s_{xz} \neq 0\}} |s_{xz}|} \tag{5.2}$$

The strategies of attackers are as follows:

1. Find the popular contents of the website. Due to the frequency of ratings on the website generally follow the power law distribution. So a small number of popular resources have been rated a lot times. These kinds of resources are generally displayed in a prominent position of the webpage.
2. Create a certain number of user accounts in batch, and mark the popular resources with the average rating of the item. In this way, the attacker's accounts have higher similarity with many actual user accounts in the system.
3. Use the above accounts to rate the target item. Due to the attacker's account have higher similarity with many users' in the system, the attacker's rating determined the output of the predicting formula.

Although commercial website usually limit the number of registered account for a single IP, it is difficult to limit dynamic IP strictly. At the same time, attackers can easily change its IP in order to bypass this limitation using P2P agent software, such as Tor.

It should be noted that, there are many types of collaborative filtering algorithms, and there are different attack methods for different approaches. For example, cosine similarity-based algorithms are less affected than the typical examples [7]. But PLSI model demonstrated strong robustness in experiment [14]. However, due to the collaborative filtering technology must rely on the similarity of user's rating records, and the reliability of rating can't be judged on comment itself, so collaborative filtering method can't avoid the attack described above completely. In addition, there are some effective work have been done by some researcher to improve the robustness of collaborative filtering algorithm [10, 12, 13]. But just as mentioned in [14], all the new method has limitations. First, the assumptions they supposed are not entirely consistent with real life. Second, these kinds of approaches need an offline process and the process is always time consuming, which limits their scalability.

Finally, those methods need to set threshold to filter noisy data manually, which may lead to false judgments or low prediction coverage.

Due to the reliability of the information can't simply be judged from the format and text features, there are limitations on robustness of collaborative filtering. We need some ways to measure the reliability of information source, and then asses the reliability of information by itself. Because different users have different opinions, measurement should be done from the point of target user. The results of the assessment will be weighted and then generate comprehensive prediction. The higher the trust degree of a user, the higher proportion his/her rating is in the collective view predicting results.

In order to achieve the trust-based collective view prediction, two basic elements are needed: trust data and prediction algorithm. Trust data refers to the mutual evaluation between users. This evaluation may be given directly by users through trust statement, or can be analyzed based on the collection of users' interaction records. Trust network, a special form of social network, can be built by regarding users as nodes, and supposing trust relationships between users as edge. It is a main source for calculating the user trust degree. The input of the algorithm is the target user u_x, item i_y which is to be predicted, the user collection U_y which have reviews on item i_y and the trust network G. The algorithm gives each user in U_y an appropriate weight based on the relationship between u_x and U_y in G, and then provide the final weighted results. Different from the collaborative filtering algorithm which find the association between user's opinions directly, trust-based collective view prediction method captures the association between opinions and the relationships of their authors in trust network, according to which it gives weights to different authors for a particular target user. Therefore, the data and the algorithm are closely contacted with each other. The design of prediction algorithm should be based on the hidden correlation between the trust network and user insights.

In the following of this chapter, we will first define the application environment of group insights prediction, and the conceptual difference and connection between it and user similarity, as well as the method to build trust network. And then, we will introduce the statistic feature of it, and the principles to calculate the trust degree between the indirectly connected nodes in trust network. On this basis, we use the data from famous review website Epinion.com and Chinese website DianPing.com as sample to analyze the impact of the node distances in network, and trust degree to the user opinion similarity.

5.2 Trust and Its Measurement

Trust is a broad sociological concept. For example, "trust is the attitude towards risk [150]"; "trust is the subjective expectation in the interest of individual A to the specific behavior of individual B [151]"; "trust refers to a person willing to rely on other people or things in certain circumstances, he/she can feel relatively sage, despite the negative result is still likely to occur [23]". Some scholars try to give a more detailed definition. McKnight [152] defined trust as three levels, namely, "trust behavior", "trust will" and "trusted sources". The trusted source is composed of "system trust", "context trust", "character trust" and "belief sources". It generates trust will and eventually leads to trust behavior. Grandison [98] who focus on Internet environment, presented trust as five dimensions, namely, "identity trust", "provider trust", "visitor trust", "context trust" and "proxy trust". Due to the broad notion of trust, when we use the term, we must limit its specific contents, or it will lead to the concept as a "very general, but no avail concept, as it will include almost all of the things around us [151]".

According to the research object of this study, we use the definitions which are similar with that in literature [17, 20].

Definition 1 Trust is the belief of subject user x to object user z who is able to reduce the uncertainty of unknown things. The strength of belief is called trust degree.

Specifically, the subject user x is the person who wants to know about some unknown item y, such as goods, restaurants and so no. The object user y is the person who has wrote comments about item y. If x trusts z, z's comments are hopeful to x which can reduce the uncertainty of item y. Different subject users have different ideas about the ability to the same object user. This kind of trust conflicts (i.e., some user trust a comments author, but at the same time some users does not trust him) has been found in the study on Epinions.com [153]. Therefore, trust is a subjective judgment.

According to the above definition, there are essential differences between trust and user similarity, but they also have some correlations. In a broad sense, the user similarity can be used as one of the sources to build trust. Assume that user x and user z have similar comments on the items that they have reviewed, z's insights about item y which is unknown to user x is helpful for user x to reduce the uncertainty of item y. And of course it is the same in reverse. The higher the similarity, the stronger ability is to reduce the uncertainty. From this perspective, it is reasonable for studies [18–20] to define user similarity as trust. However, we believe that the user similarity is only one of the sources to build user trust relationship, and whether the source will be used to build depends on the specific application environment. In a completely opening review website, the attackers can take advantage of user similarity to achieve their own evil purpose. So it is unreasonable to use user similarity as a source of trust. But the user structure is relatively homogeneous in some application environment, such as academic resources recommendation system in university and research departments. User similarity can be used to build trust in such context, for the probability of those users to attack is low.

5.2.1 Data Source

How to build trust in Web environment? In review system, user is not only the commenter but also the content creator. The comments are not only on commented items (products or services) but also on comments. Figure 5.2a shows that user x can comment on the contents which are created by user z after reading them. This evaluation reflects the former's recognition of the latter's capacity, which is called trust. In addition, the user x can update the trust relationship with user z directly based on the experience of user x. These are different from collaborative filtering. As shown in Fig. 5.2b, the only status of user is evaluator in collaborative filtering. The type of relationship is similarity, which reflects the relevance of user's ratings. Another significant difference is that the trust relationship is asymmetric. The evaluation of user x to user z only impacts the trust degree of former to latter. But the user similarity is generally symmetric, which leads to the vulnerability of collaborative filtering.

The review site such as Epinions.com provides the trust management function (Fig. 5.3). The user can declare trust or shield (do not trust) an author, thus affecting the browsing experience of themselves and the users who trust them [154]. Epinions.com provides guidance of how to declare trust statement: (1) Read all or most of the reviews. (2) Check their basic information. (3) Assess their trust network. In some sites, such as KouBei.com and DianPing.com, the similar function is provided by "friend list". When we use trust function to get trust degree, the input is the rating of user x on the comments which are created by user z. And then the trust degree is calculated by weighted formula.

The two kinds of trust relations have their own advantages and disadvantages. The mode of trust statement gives users greater control power. When a new user join the virtual community, she can choose trusted partners in this way, for example, trust a friend who send out the invitation, thus take the trust relationships from the real world or other systems to the new virtual community. The disadvantage is that the update of the trust relationship is completely maintained by the

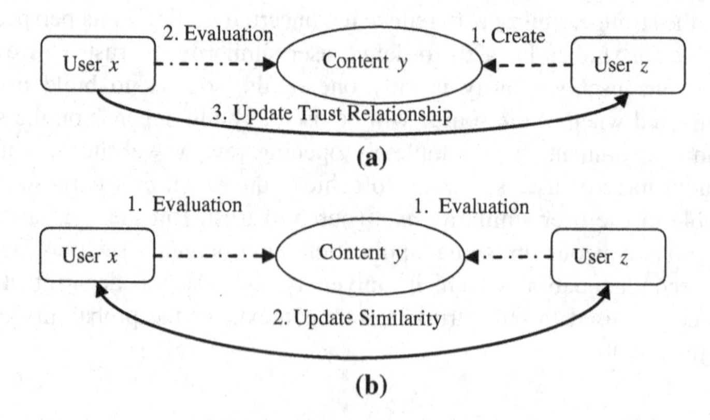

Fig. 5.2 Update mode of trust degree and similarity

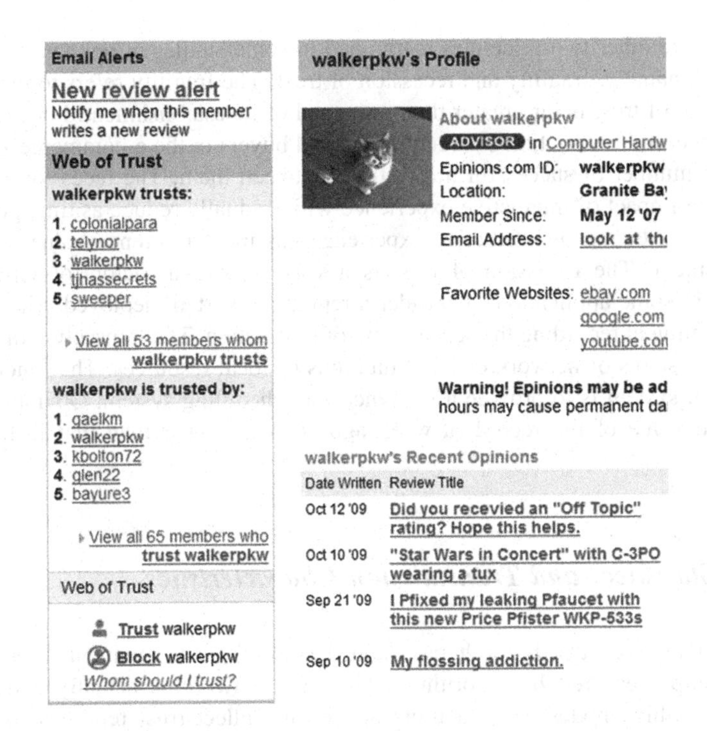

Fig. 5.3 The trust management functions at Epinions.com

user. Some users may join an author to the trust list only by reading one comment of his/her, and some users may do this after reading all the author's reviews. These differences in behavior may have bad impact on closer neighbor selection process which based on trust network. In the latter way, the user only need to comment on his/her interest reviews, and the update of trust relationship is controlled by the system. Obviously, it is much easier to rating on a comment than to decide whether to trust a person. The disadvantage in the mode is that the update process is a slow one. Users need to interact repeatedly before significant changes in trust happened, so it can't take the prior knowledge (e.g., friends) of the user into consider. Therefore, a more reasonable measure is to combine the two methods, and use these two types of technology flexibly.

The design of trust function should follow two basic principles. First of all, trust is a gradual process of evolution. Only when the user x repeatedly praised to the user z, the former have a higher trust degree on the latter. Secondly, the expression of trust degree should reflect the degree of certainty of the results. Assume that user x have evaluated 10 times to user z, and the times of evaluation of user x to user y is 5, and then the certainty which calculated based on 10 times rating is higher than that based on 5. Trust measurement must take the above principles into account, for we need to merge the results of many paths in trust network. If we only use a real number to reflect trust degree, some useful information may lose.

There are other two principles mentioned in some studies of reputation system [23, 101], namely, fragility and recession of trust. The fragility refers to the speed of collapse of trust is far greater than the speed of its establishment. For example, an unsuccessful transaction between sellers and buyers in the e-commerce site can denied a number of successful transactions between them. The recession of trust means the impact of interactive experience will gradually reduce as time goes by. In other words, recent interactive experience has more reference value for trust measurement. The recession of trust is a very important factor for reputation systems in some applications. Consider a reputation system deployed grid system [155], aiming at recording the success rate of transaction. The transaction may fail, due to the status of network or the limitations of local resources. The function of reputation system is to provide a reference for scheduling system. Obviously, the reference value of the records a week ago is much lower than that in the day before.

5.2.2 Statistical and Transmission Characteristics

The trust network can be built based on the users' trust statement or the trust relationship measured by algorithms. The trust network is actually a directed graph, in which nodes refer to users and edges reflect trust relationship. As a special form of social network, trust networks have the characteristics of social network, such as power-law distribution, high connectivity, small-diameter, and the clustering effect. Some characteristics have an important effect on the prediction on trust network. The high connectivity ensures that we can find the path between two nodes with a high probability. The small-diameter character makes the search process can be completed with a few paths. Power-law distribution means there are a small number of nodes in the trust network with high in-degree. The insights of such nodes have a greater impact on the output of trust-based prediction algorithm. Therefore, when we analysis the robustness of the prediction algorithm, we should pay attention to that attacker may take advantage of the above features.

In addition, the majorities in the community do not have interactive experiences between each other under normal circumstances, so there is no trust relationship data provided to the system. It is difficult to predict directly based on the limited direct trust relationship, since the predicting algorithm can't guarantee to have all the needed information to do prediction (i.e., the direct trust relationship between the target user u_x and the user collection U_y who have commented on item the majorities in the community). To solve the data sparse problem, we should calculate the trust degree of the indirect adjacent nodes based on the trust transmission feature.

The propagation characteristics of the trust, including transitivity and composition [17], correspond to two cases in Fig. 5.4. Based on these two features, we

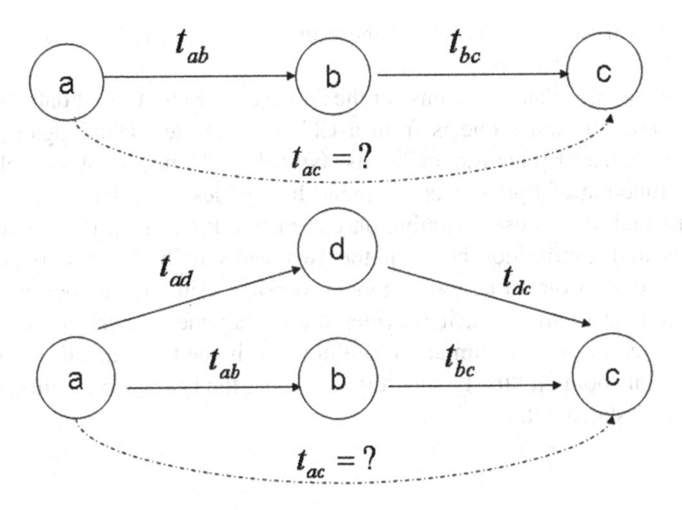

Fig. 5.4 Diagram of the transmission and composition of trust

can merge the calculated values from many trust paths and then get the final measurement results. It should be noted that the specific meaning of transitivity and composition of trust has a big difference depend on different methods of measurement. Ziegler [16] has made a classification and induction of trust, but he did not give an absolute standard to evaluate which strategy is the best. We believe that the choice of different trust measurement methods is scenario-oriented. In the next chapter, we will design prediction algorithms based on different trust measurement method, and also analysis their differences.

5.3 Analysis of Trust and Collective View

The premise to design reasonable collective view prediction model is to discover the inherent laws of trust network. Specifically, there must be some correlations between the structure of trust network and the similarity of user's opinion. We can give different weights on different opinions according to the above correlations, and then complete the prediction.

There have been some researches on the relationship between trust and user similarity in Web. For example, people tend to make friends with user who have similar interesting in virtual community [156, 157]. Ziegler [16] found that the similarity of reading behavior between user and the users in the trust list if much higher than others who are not in the trust list in AllConsuming.com. However, all those studies analysis the relationship between similarity and direct trust of people. Due to the sparse problem of dataset, we evaluate whether the indirect node is trustable based trust metric when predicting the collective views. Is there such kind of relationship between the unconnected nodes in trust network? Are the distances of nodes influence the degree of relation? What forms of the associations between similarity and trust? The answers of these questions determined whether we can predict collective view based on trust.

We analyze three datasets to answer the above questions. Two of dataset is from Epinions.com. The other one is from a Chinese website "DianPing.com". The reason to choose Epinions.com is its typical and functional completeness. The main function of Epinions is comment. It provides the management function of trust list and allows user to rating on comments. It can help us to analyze the differences and similarities between the two ways to build trust relationship. "DianPing.com" is one of largest comment websites. Although it does not provide the management of trust list, it provides the management of friends list. A significant between Epinions.com and DianPing.com is the first one allows the users give statements both on "trust" and "distrust", and the last one only allows user to take "friend" statement.

5.3.1 Dataset

The first dataset is the crawled results for five weeks provided by Massa and Avesani [15]. We call this dataset as EPP. It contains 49,290 users, 139,738 products, and 664,824 ratings (1–5) and 487,181 trust statements. Each rating to product corresponds to a comment. The dataset does not contain distrust list, for Epinions.com have not made the distrust list public in order to avoid the malevolence between users. The data is crawled from the selected source node, and the method is bread-first-search. Therefore, when the crawling is termination the ratings and trust list of the node in the outermost is not contained in the dataset. The final experiment dataset only contain the data which appears in the trust network and rates at least one product.

The second dataset is provided by the manager of Epinions.com. We call this dataset as EPR. It contains 196,000 users, 717,000 trust statements, 124,000 distrust statements, and more than 20,000,000 ratings (from rating 1 to 6). It should be noted that the EPR dataset does not contain the ratings of products provided by users. And only a small number of users in special status (known as Advisor) can rate 6 to other's comments. We replace 5 with 6, for only a small proportion of ratings are 6. The EPR does not require data preprocessing such as EPP data, for the EPR data is obtained directly from Epinion.com. However, we found that there are some comments that missing the corresponding authors. So we remove the records related to the half-baked comments from the dataset.

The third dataset is crawled from a Chinese website "DianPing.com". We call this data set as "DP". The crawling process is from January 24, 2010 to March 20, 2010. Due to the limitation by DianPing.com, we send a HTTP request per 10 s. The crawling is beginning from a randomly selected initial node, and then the breadth-first search is used to find the friends lists, finally the node information is crawled recursively. We remove the related information of the users whose friends list is non-public information. The final dataset contains 58,109 users, 175,215

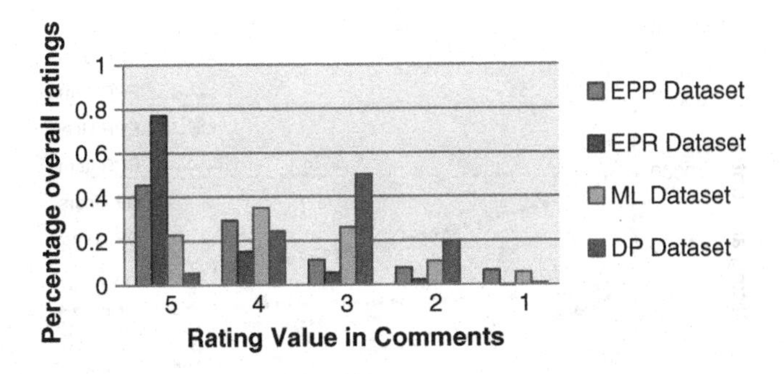

Fig. 5.5 Rating distribution of the data

items, 1,184,514 ratings (from 1 to 5), and 91,577 friend statements. Some comments in DianPing.com contain a comprehensive rating, ranging from {very poor, poor, not bad, good, and very good}. We give the 1–5 ratings correspondingly.

Figure 5.5 shows the rating distribution of the data. We use the MovieLens (ML) dataset as the baseline. The ML dataset is collected from the MovieLens web site. A lot of collaborative filtering research use ML as the experimental data set, for it is very representative. Figure 5.5 shows that the ratings from EPP data set are relatively uniform, but it still biases high rating significantly, more than 75 % ratings are 5. In contrast, the data from DP are obeying normal distribution and the mean is 3. ML data are obeying normal distribution and the mean is 4. Additionally, the sparse problem of the data is significantly different. The definition of sparse rate is the ratio of the empty elements in the user-item rating matrix. The sparse rate of EPP data is 99.988 %, EPR data is 99.985 %, DP is 99.979 % and the ML is 95.532 %.

Figure 5.6 shows the distribution of the numbers of comments provided by the top 5,000 active users, and the number of comments on the top 5,000 popular items. Figure 5.7 shows the distribution of the top 1,000 in- and out-degree of node in trust network. It can be seen that the number of users' ratings, the number of comments of items, and the in- and out-degree of node in trust network all follow law distribution.

5.3.2 *Measurement*

5.3.2.1 Similarity Measure

We use Cosine and Mean Absolute Deviation (*dev*) to measure the similarity of user's opinion. The two measure methods focus on different aspects of similarity. Cosine similarity reflects the degree of interests of the user to items. The definition is as follows:

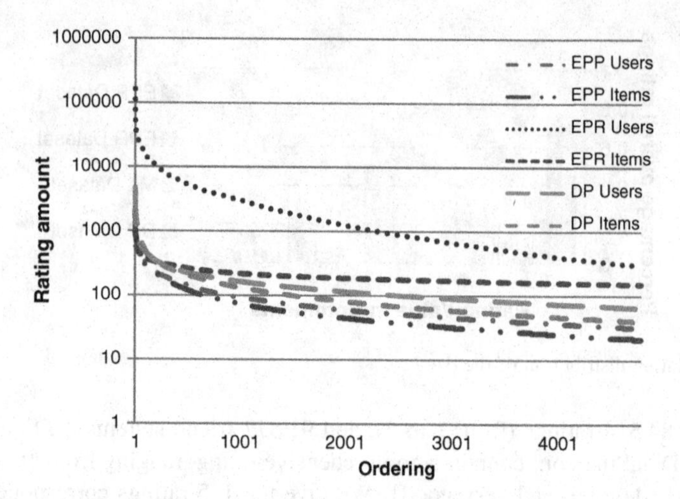

Fig. 5.6 Comparison of rating distribution among different datasets

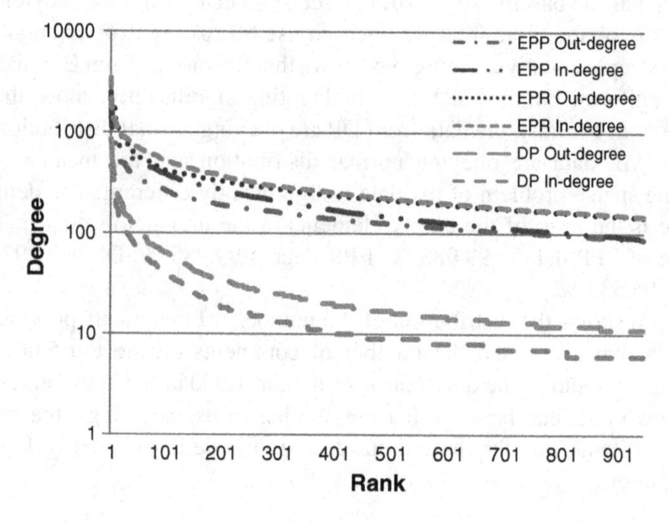

Fig. 5.7 The distribution of in- and out-degree of node in trust network

$$s_{xz} = \frac{\sum\limits_{\{y|\rho_{xy},\, \rho_{zy} \neq \phi\}} \rho_{xy} \cdot \rho_{zy}}{\sqrt{\sum\limits_{\{y|\rho_{xy} \neq \phi\}} {\rho_{xy}}^2 \cdot \sum\limits_{\{y|\rho_{zy} \neq \phi\}} {\rho_{zy}}^2}} \tag{5.3}$$

where ρ_{xy} is the rating user x gives to item y. The higher the Cosine similarity of two users, the higher the ratio of the common rating items to the total number of ratings of them. Due to Cosine similarity is insufficient to reflect the differences in the user's particular opinions; we also introduce *dev* to measure the similarity of users' opinions.

$$dev_{xz} = \frac{\sum_{\{y|\rho_{xy},\,\rho_{zy} \neq \phi\}} |\rho_{xy} - \rho_{zy}|}{|\{y|\rho_{xy},\, \rho_{zy} \neq \phi\}|} \tag{5.4}$$

5.3.2.2 Trust Statement and Trust Function

In the EPP data set, the trust statement is a single-value, for the distrust list is invisible. In this data set, the trust statement is either 1 or -1, presents trust and distrust respectively. We choose the Beta probability density [103] function as the trust function in EPR data set. And then build a trust network based on users' ratings on comments. Beta probability density function can reflect trust as a gradual evolution process. And the trust reflected by Beta function can be measured by trust theory [104]. It has been used in reputation system such as [155, 158], and it can reflect the fragility and recession of trust by introducing penalty factor and time decay function.

Beta distribution function is suitable to express the posteriori probability distribution of binomial events. Let's consider the simplest situation, user u gives only two rating: {positive, negative} to the user z's comments, and then the trust degree can reflect the expectation of positive rating. Trust function expresses the form of expectations of Beta probability density function for positive rating.

$$t_{xz} := \int_0^1 \frac{\Gamma(p_{xz} + n_{xz} + 2)}{\Gamma(p_{xz} + 1)\Gamma(n_{xz} + 1)} t^{p_{xz}} (1 - t)^{n_{xz}}\, dt = \frac{p_{xz} + 1}{p_{xz} + n_{xz} + 2} \tag{5.5}$$
$$0 \leq t \leq 1,\ p_{xz} \geq 0,\ n_{xz} \geq 0$$

where t is the random variable of binominal event. A_{xz} is z's comment set which has been read by user x. $\kappa(c)$ reflects the rating of user x to a particular comment. The parameter p_{xz} and n_{xz} respectively denote the number of positive and negative ratings in the set. If user x has not rated any comments, the $\kappa(c) = \Phi$. We use two-tuple (p_{xz}, n_{xz}) to denote trust, for they can reflect all the information to calculate the degree of trust. A demonstration of Beta function with different parameters is shown in Fig. 5.8. Probability density function becomes sharp with the increase of positive ratings. And expectation is closing to 1 at the same time.

The above formula applies to the case that range of the rating function is not binary. If the range of rating function is discrete interval value with global order, we could calculate p_{xz} and n_{xz} according to Eq. (5.6), where $lb(R)$ and $ub(R)$ represents the upper bound and lower bound of the possible rating values.

$$\begin{cases} p_{xz} = \dfrac{\sum_{c \in A_{xz}} \frac{\kappa(c) - lb(R)}{ub(R) - lb(R)}}{|A_{xz}|} \\ n_{xz} = 1 - p_{xz} \end{cases} \tag{5.6}$$

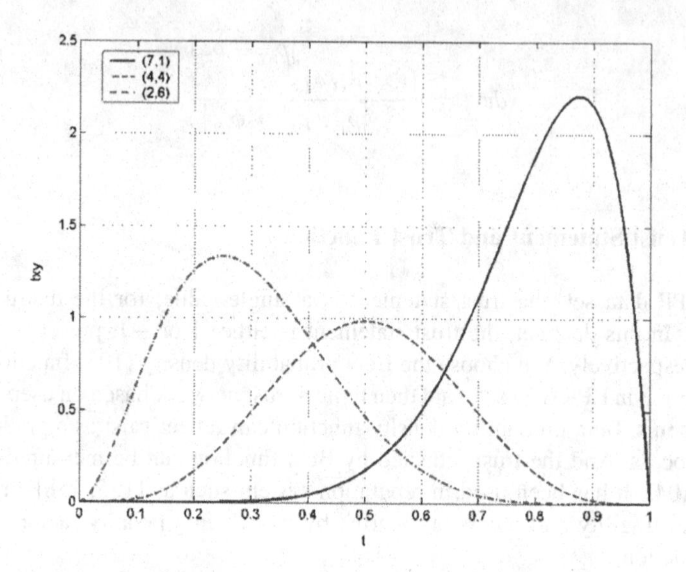

Fig. 5.8 Beta probability density function with different (p, n) parameters

As mentioned in [158], considering the fragility of trust, we can treat a low rating as many negative ratings in order to significantly reduce the input of trust function. Considering the recession of trust, we can assign a time decay factor to each rating. However, the two methods need to set parameters artificially. There are not universal rules to guide how to set parameters. Therefore, we only use Beta trust function to build trust network in the following analysis. The following analysis and experimental results show that the degree of matching is high between trust statements and the trust network built by Beta trust function. So the above assumption is reasonable.

5.3.2.3 Trust Inference

Due to the data sparse problem in real web systems, direct trust metric usually can only be calculated between small parts of users. To improve the coverage of trust metric, we need to rely on indirect trust metric (i.e., inferring trust over the trust web). Trust inference is based on two widely accepted properties of trust, namely trust decay transitivity and composition. Breadth-first search is a widely used trust inference strategy in the literature [15, 17, 19, 20]. Although these strategies have many differences in detail, the basic principle is similar to MoleTrust [15] approach. Here, we take MoleTrust as an example to explain the principle. As shown in Fig. 5.9, assume the task is to calculate the trust from source node a to all reachable nodes. First, the trust network can be converted into a directed acyclic graph taking node a as the center, and delete the edges connecting the nodes with the same degree. Then, starting from a node whose depth is 2, calculating indirect

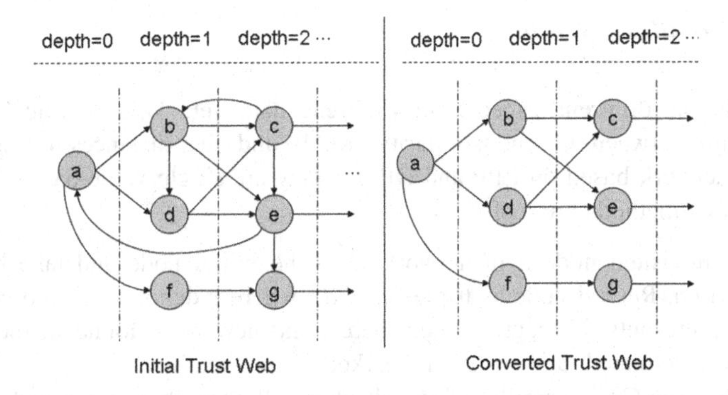

Fig. 5.9 Example of MoleTrust

trust ω_{ac} by using the Eq. (5.7) repeatedly, until no new nodes can be calculated. Where TT indicates the threshold of trust, $B\ c$ represents a set of nodes which points to node c in the converted graph.

$$\omega_{ac} = \frac{\sum\limits_{\{w|u_w \in B_c \wedge \omega_{aw} > TT\}} \omega_{aw} \cdot t_{xc}}{\sum\limits_{\{w|u_w \in B_c \wedge \omega_{aw} > TT\}} \omega_{aw}} \tag{5.7}$$

A problem with MoleTrust is that the converted trust web may lose some reachable nodes on the initial web. As shown in Fig. 5.9, if $\omega_{af} \leq TT$, node g will be unreachable on the converted trust web. However, if the inferred trust $\omega_{ae} > TT$, we would look forward to calculate $\omega_{ae} > TT$ via node e. In order to overcome the problem of MoleTrust, we use a progressive breadth-first-search algorithm as our trust inference method. The pseudo codes for inferring trust from the source node s to the target node t are shown in Appendix 1.1.

BetaTrust follows the similar idea as MoleTrust. In order to combine the trust represented as beta distribution, it introduces two operators from Belief Theory [104]. The operator $\otimes$ is used to discount trust along the inference path (see Eq. 5.8), and the operator $\oplus$ is used to combine the results from different inference paths (see Eq. 5.9). The above operators are used to replace the weighted average computation defined in Eq. (5.7). The trust inference algorithm is similar with the one with real value (see Appendix 1.2).

$$\omega_{ac}^{b} = t_{ab} \otimes t_{bc} = \left(p_{ac}^{b}, n_{ac}^{b}\right)$$
$$\begin{cases} p_{ac}^{b} = \frac{2p_{ab}p_{bc}}{(n_{ab}+2)(p_{bc}+n_{bc}+2)+2p_{ab}} \\ n_{ac}^{b} = \frac{2p_{ab}n_{bc}}{(n_{ab}+2)(p_{bc}+n_{bc}+2)+2p_{ab}} \end{cases} \tag{5.8}$$

$$\omega_{ac} = \omega_{ac}^{d} \oplus \omega_{ac}^{b} = \left(p_{ac}^{d} + p_{ac}^{b}, n_{ac}^{d} + n_{ac}^{b}\right) \tag{5.9}$$

5.3.3 Indicator Systems

We focus on different aspects on different data set. We will analyze the relationship between Cosine similarity, MAE and the distances of nodes in the trust network based on EPP and DP, for they are single value-based data set. The analysis methods are as follows:

1. Set an unvisited node u in network G as the initial node and take breadth-first-search. Record node as triple (u, v, d), where v denotes the current node, and d represents the depth. When there is no new node founded, the search process terminated, and node u is marked as "visited".
2. Calculate the Cosine similarity and absolute difference between u and v for all the triple (u, v, d), and then denote the results as (s_{uv}, d) and (dev_{uv}, d). If there are no common rating between u and v, the value of s_{uv} is 0, and dev_{uv} is not recorded.
3. Calculate the average value of $(\bar{s}_u, d)$ and $(\overline{dev}_u, d)$ for all (s_{uv}, d) and (dev_{uv}, d) got in the same depth.
4. Repeat the above process for all the nodes that have not been visited in graph G. If all the nodes are visited, calculating is terminated.

In order to analyze the influence of d for user similarity, we take the average of all $(\bar{s}_u, d)$ and $(\overline{dev}_u, d)$ for each depth. As the benchmark for comparison, we define the "benchmark Cosine similarity" as the average similarity of node-pair in G, and we also define the "benchmark absolute deviation" as the average deviation between all the node pair.

A binary trust network and trust network whose weight is the expectation of Beta distribution can be constructed based on EPR data set. Just as shown in Fig. 5.4, most of the rating is 5 in EPR data set, so the average differences of users' ratings is very small. Therefore, we only analyze the relationship between Cosine similarity and other indicators. We believe that two users' rating on items will be similar if the similarity between them are high, for user's rating on comment represent the cognition of the comments.

Similar to the EPP analysis, using the trust inference method discussed above, we can get four tuple (u, v, d, t_{uv}), and (s_{uv}, d, t_{uv}) corresponding, and then get $(\bar{s}_u, d, t)$. In order to use variance to do analysis, we divide trust degree into discrete intervals. In binary trust network, we use $TT = 0$ as the boundary, and divide trust degree into two intervals $[-1, 0]$ and $[0, 1]$. In Beta trust network, we use $TT = 0.5$ as the boundary, and divide the range $[0, 1]$ into 10 intervals according to the step size of 0.1.

In addition, we analysis the similarities and differences quantitatively between the two methods of building trust network on EPR data set. First, we statistic the number of nodes contained in the two trust networks. The more number of nodes in the trust network is, the more capable to predict trust measurement results for more users are. Then we select the overlap edges in the two trust network, and set the threshold of Beta trust function $T_p = 0.5$. If the Beta trust degree is higher than the threshold, the user will make the trust statement. Otherwise the user will make the mistrust statement. Based on the above method the Beta trust function is mapping to $\{1, -1\}$ in order to compare with the binary trust network.

5.3.4 Result Analysis

5.3.4.1 EPP Data Set

Figure 5.10 shows that user average Cosine similarity decreases with the increase of the distance between nodes. When the nodes distance $d \leq 4$, Cosine similarity is greater than the similarity between the randomly selected nodes. $d = 13$ is the outlier. But just as showed in Fig. 5.10, the number of $(\bar{s}_u, d)$ can be observed is only [114], so the outlier is the impact of noise data. The statistical significance analysis result is showed in Table 5.1. We examine the impact of node distance d to the observed variables $(\bar{s}_u, d)$, adopting single-factor variance analysis. It can be seen from Fig. 5.11, due to the small diameter characteristics of the network, the average number of observed $(\bar{s}_u, d)$ when $d > 10$ is significantly less than that when $d \leq 10$ in EPP trust network. So we only consider the case when in $d \leq 10$

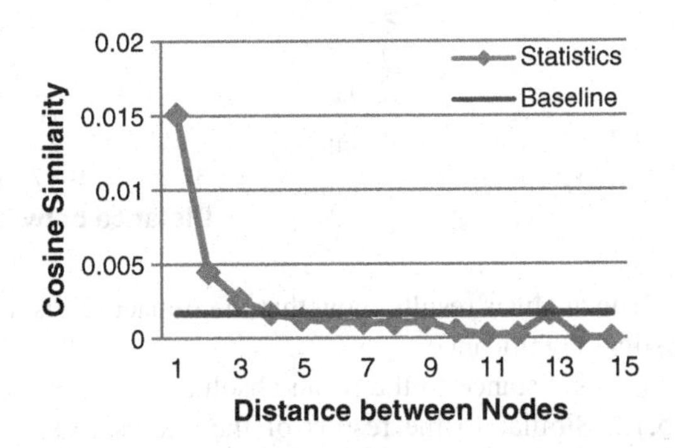

Fig. 5.10 The relationship between user average cosine similarity and the distance between nodes in trust network on EPP dataset

Table 5.1 The variance analysis result of the relationship between user average Cosine similarity and the distance between nodes in trust network

The sources of variance	Sum-of-squares	Degree of freedom	F value	Critical value	Significance
Between group	4.75	9	3,120.36	$F_{0.05}(9, \infty) = 2.71$	Very significant
In group	44.50	263,284			
Sum	49.25	263,293			

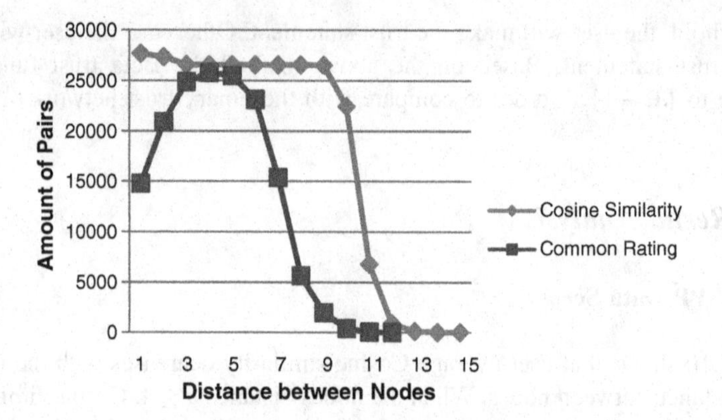

Fig. 5.11 The relationship between observed number of $(\bar{s}_u, d)$ and $(\overline{dev_u}, d)$ in EPP dataset and the distance between nodes in trust network

Fig. 5.12 The relationship between mean absolute deviation (*MAD*) of user ratings in EPP dataset and the distance between nodes in trust network

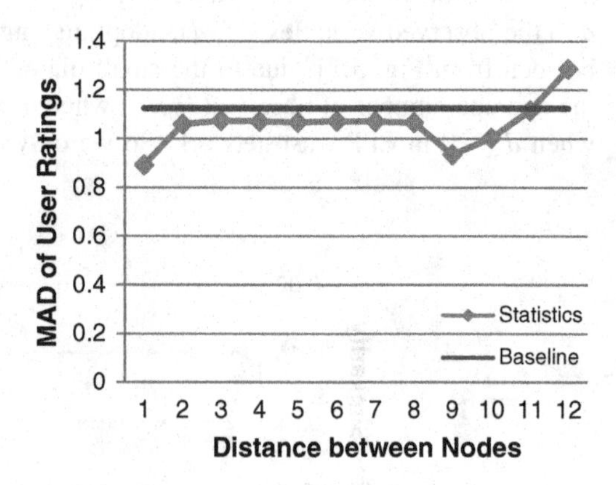

variance analysis. The analysis results show that the impact of distance d to Cosine similarity is statistical significance.

The impact of nodes distance to the mean absolute deviation of user ratings is showed in Fig. 5.12. Similar to the results of the analysis of Cosine similarity discussed above, the mean absolute deviation of user ratings increased with the increasing of distance d. $d = 9$ is the outlier. Since the number of observed $(\overline{dev_u}, d)$ is small, the outlier can be considered as the impact of noise data. The results of variance analysis based on $d \leq 6$ are showed in Table 5.2. The results show that the impact of nodes distance d to the mean absolute deviation of user ratings is statistical significance.

Base on the analysis of EPP dataset, we found that the single value trust network contains the information of the similarity of user opinions. And the closer the distance between nodes is, the stronger the similarity will be.

Table 5.2 Variance analysis results of the relationship between the mean absolute deviation of user ratings and the distance between nodes in trust network

The sources of variance	Sum-of-squares	Degree of freedom	F value	Critical value	Significance
Between group	414.02	5	271.05	$F_{0.05}(5, \infty) = 4.36$	Very significant
In group	41,204.44	134,879			
Sum	41,618.46	134,884			

5.3.4.2 DP Dataset

It can be seen from Fig. 5.13, similar to the results on EPP dataset, the Cosine similarity between users decreased with the increasing of distance between nodes. When the distance is less than 5, the Cosine similarity is greater than the similarity of randomly selected nodes. In addition, the characteristic of small diameter is less significance in DP dataset than that in EPP data set, which can be seen from Figs. 5.13 and 5.14. This may be caused from that we excluded those who did not

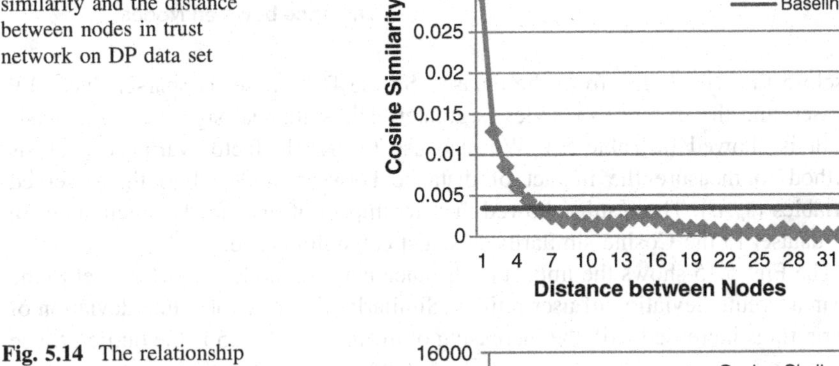

Fig. 5.13 The relationship between user average cosine similarity and the distance between nodes in trust network on DP data set

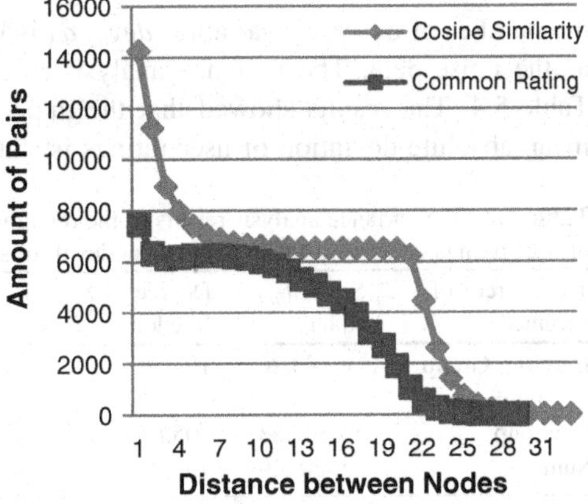

Fig. 5.14 The relationship between the number of observed $(\bar{s}_u, d)$ and $(\overline{dev_u}, d)$ and the distance between nodes in the trust network

Table 5.3 The variance analysis result of the relationship between user average Cosine similarity and the distance between nodes in trust network

The sources of variance	Sum-of-squares	Degree of freedom	F value	Critical value	Significance
Between group	10.45	19	1,140.49	$F_{0.05}(19, \infty) = 1.88$	Very significant
In group	71.81	148,915			
Sum	82.26	148,934			

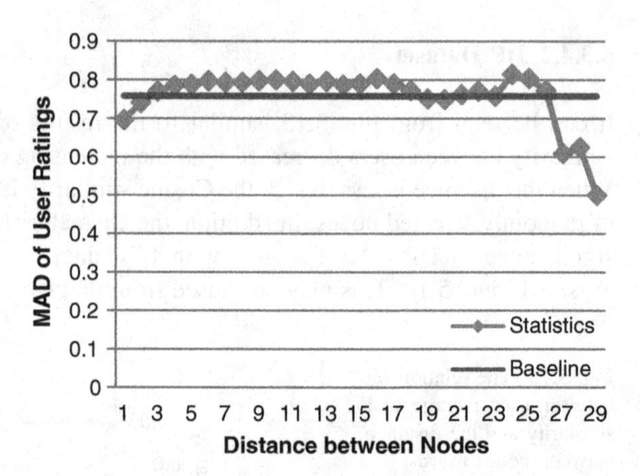

Fig. 5.15 The relationship between mean absolute deviation of user ratings and the distance between nodes in trust network

disclose the friends list from the dataset. So the DP dataset is sparser than EPP dataset, and the distance of nodes is greater. The statistical significance analysis result is showed in Table 5.3. We also use the single factor variance analysis method to measure the impact of distance between nodes d to the observed variables $(\bar{s}_u, d)$. The results showed that the impact of distance between nodes in DP dataset to the Cosine similarity is statistical significance.

The Fig. 5.15 shows the impact of distance between nodes in DP dataset to the mean absolute deviation of user ratings. Similarly, the mean absolute deviation of user ratings increased with the increasing of distance d. $d \geq 25$ is the outlier. Since the number of observed variable $(\overline{dev_u}, d)$ is less when $d \geq 25$, it can be consider as the noise data. The variance analysis results based on $d \leq 20$ are showed in Table 5.4. The results showed that the impact of distance between nodes to the mean absolute deviation of user ratings is statistical significance.

Table 5.4 The variance analysis results of the relationship between the mean absolute deviation of user ratings and the distance between nodes in trust network

The sources of variance	Sum-of-squares	Degree of freedom	F value	Critical value	Significance
Between Group	83.26	19	30.1958	$F_{0.05}(19, \infty) = 1.88$	Very significant
In Group	15,365.23	105,882			
Sum	15,448.49	105,901			

It can be seen from Table 5.4, the results of mean absolute deviation of ratings on DP dataset is very different from that on EPP dataset. The mean absolute deviation of ratings between connected nodes in latter's trust network is less than the reference value. But in the former's trust network, the mean absolute deviation of ratings between connected nodes is greater than the reference value when $d \geq 3$.

5.3.4.3 EPR Dataset

Since EPR dataset contains more users than EPP dataset, in order to reduce the computational complexity, we select 1,000 common nodes in the two trust network based on random sampling method to do statistical analysis. The relationships between Beta trust degree, Cosine similarity and the distance between nodes are showed in Fig. 5.16. It can be seen from that, the Cosine similarity is positively correlated with Beta similarity, and negatively correlated with the node distance d. When $d \geq 3$, the impact of trust to similarity is dramatically decreased. $d = 1$, $t = 0.2$ is the outlier. But the observed number of $(\bar{s}_u, d, t)$ is only 2, so it can be considered as noise data.

The relationships between binary trust degree, Cosine similarity and node distance are showed in Fig. 5.17. It can be seen from that, similar to Beta trust network, the impact of trust degree and node distance to similarity is also statistical significance. When the distance is more than 3, the impact of trust degree to similarity weakened gradually (Fig. 5.17).

The comparison chart of number of node in Beta trust network and binary trust network is shown in Fig. 5.18. It can be seen from that, the Beta trust network contains more user nodes, so it can predict collective views based on measuring trust degree for more users. The correlation analysis results of Beta trust network and binary trust network are showed in Table 5.5. Among them, the definition of

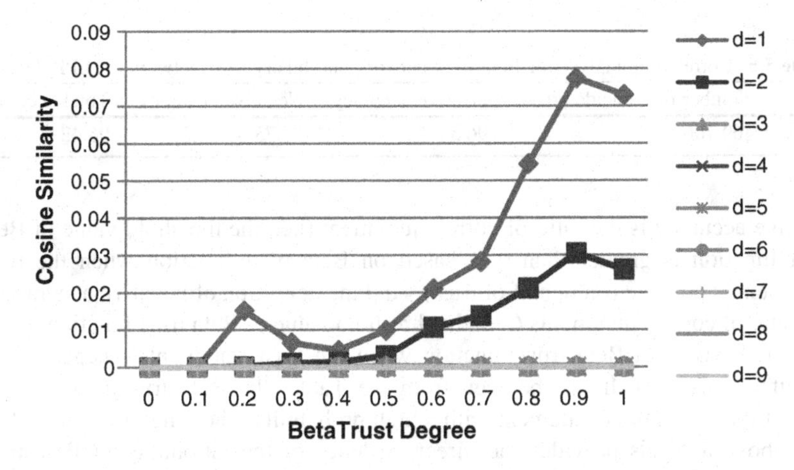

Fig. 5.16 Relationships between Beta trust degree, Cosine similarity and node distance d

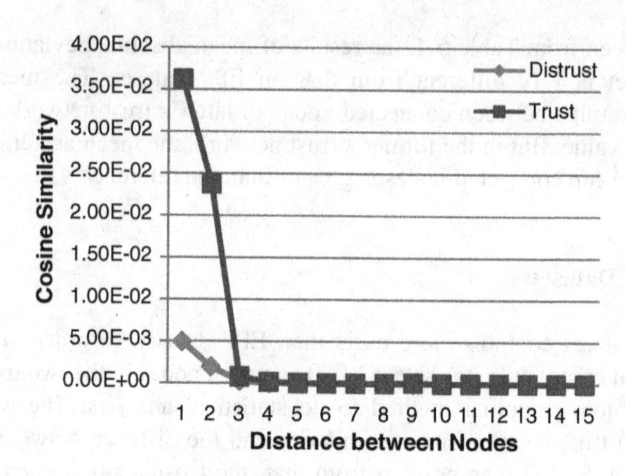

Fig. 5.17 Relationships between binary trust degree, Cosine similarity and node distance in EPR dataset

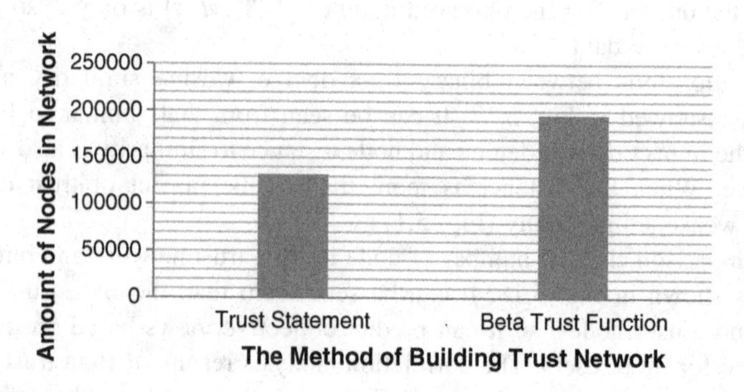

Fig. 5.18 The impact of different method to build trust network to the number of nodes in network on EPR dataset

Table 5.5 Correlation analysis of Beta trust network and binary trust network on EPP dataset

	Number of coincide edges	Positive accuracy	Negative accuracy	Total accuracy
Value	469,766	96.34 %	97.75 %	96.42 %

positive accuracy is the ratio of correct judgment (i.e., the threshold value of Beta trust function is greater than 0.5) based on Beta trust function when the trust statement is positive (on behalf of trust). And the definition of negative accuracy is the ratio of correct judgments (i.e., the threshold value of Beta trust function is less than 0.5) based on Beta trust function when the trust statement is negative (on behalf of mistrust). It can be seen from the Table, the Beta trust function can predict the user's trust statement with a high probability when the threshold is 0.5. The above analysis provides the direct evidence of the rationality of Beta trust function.

5.4 Summary

The traditional collective views predicting methods, including content-based recommendation and collaborative filtering, lack the method to measure the reliability of sources of information. So it can be attacked by rival easily. Trust network is a kind of directed graph, and can be used to improve the vulnerability of neighbor selection strategy which based only on similarity. However, in order to product good prediction results, there must be some relationship between the structure of trust network and similarity of collective views. In this section, we have defined the concepts of difference and correlation between trust and similarity in the collective views prediction context. And we also expand the quantitative analysis of their relationships. The main conclusions are as follows:

First, the similarity between nodes decreased with the increasing of nodes distance in trust network, and it correlated with trust positively. Therefore, it is reasonable to build prediction algorithm using breadth-first search to select closer neighbor, and then predict rating result using trust degree as weight. In addition, we also found that, the statistical analysis results in Epinions.com dataset and DianPing.com are very different. The former dataset shows a more obvious "small world" network characteristic. But in the later dataset, the characteristic of "small world" is not showed obviously. And the mean absolute deviation between adjacent nodes is generally greater than that between randomly selected nodes when distance is greater than 3. We believe that there are two main reasons causing the difference. First, some of the users in DianPing.com did not disclose their trust list, so we remove them from the dataset. In this way, the final trust network became sparser due to losing some original information, and then the distances between nodes are increased. Second, there is no clear guidance for user to build there "friend" list in DianPing.com. We found that some users did not make any statements, but they also became friends of some other users. Some users add the others who asked for being friend with them to the friend list just out of courtesy. But in Epinions.com, there is clear guidance to tell users they should be cautious when making trust statement. We can conclude that, in order to make better use of trust measurement and user similarity, the website should allow users to understand the role of trust statement and guide them to manage trust list carefully.

Secondly, using users' feedback to comments as input to build trust network based on trust function is able to reflect the trust relationship between users accurately. Moreover, in the case of the same user scale, compared to build network based user statement, the above method can contain more user nodes and can improve the coverage of trust network. In addition, although we can't build Beta trust network through crawling, but there is reason to believe that compared to build network based on "friends", this method have stronger correlation with the similarity of collective views. For in the "friends" mode, user can make trust statement only out of courtesy, bringing more noise data.

In the next chapter, we will explore to build collective view prediction algorithm using trust measurement based on the analysis results of this chapter.

Chapter 6
Models, Methods and Algorithms

In the previous chapter, we get the inner connection between trust and users' view by statistical analysis. The result shows that, users' similarity is positive correlated to their trust degree and closeness in trust network. In this chapter, we make a further study on how to build algorithms of collective view prediction by this law

To begin with, we explore the difference between traditional learning algorithms by examples and prediction algorithms based on trust. Then, we analyze the adaptability problems of two representative trust metrics in literature on building collective view prediction algorithms, and refine the limitations of the metrics. Then we proposed two kinds of collective view prediction algorithms based on instance-based learning by the refined trust metrics methods. Furthermore, we come back to collaborative filtering prediction completely based on users' similarity, and solve the data sparse problems by the transmission of similarity. Finally we explore to combine several predictors by Bayesian theory to further improve the accuracy of prediction.

The algorithm proposed by this chapter can be adapted to different application environment. In an open system, due to the existence of attackers, we can effectively reduce the risk of attack by using trust metrics as the main evidence to choose neighborhood for prediction algorithms. In a relatively close application environment, the similarity-based Markov model is also a practical choice. In the end of this chapter, we will make a comparison on the accuracy of different algorithms mentioned above in regular data sets.

6.1 Theoretical Models

First we give out a formal description of the problems to be solved. Given a user u_x, it can be expressed as one or two feature vector: I_x and G_x, that is $u_x = \{U_x, G_x\} \cdot I_x$ represents the rating vector of u_x, G_x represents the trust relation vector of u_x. Assumed that there are m users and n comment objects, and the trust relations between users' ratings and trust relations between users are expressed by

T. Luo et al., *Trust-Based Collective View Prediction*,
DOI: 10.1007/978-1-4614-7202-5_6, © Springer Science+Business Media New York 2013

$\mathbb{R}$. The purpose of the prediction algorithm is to get the objective function $\rho'(u_x, i_y) : \mathbb{R}^m \times \mathbb{R}^n \rightarrow V_I$ of user x and unrated object y. V_I represents the possible rating, $\mathbb{R}^m$ and $\mathbb{R}^n$ represent users' trust relation vector space and users' rating vector space respectively. From the objective function we can see that collective view prediction based on trust is richer in input compared with collaborative filtering, the latter just use users' rating records to build the objective function, that is $\rho'(u_x, i_y) : \mathbb{R}^n \rightarrow V_I$.

There are several methods to get the objective function from training methods. One method is to build a general description for objective function. In the research area of collaborative filtering, this method is called Model-based method [2, 3]. For example, in the PLSI model [159], every user and comment object is given several hidden classes according to the subordinate relation. Prediction is computed by the weighted subordinate relation. In the Model-based algorithm, the hypobook space is commonly given out and then the EM algorithm or Gradient descent algorithm is used to search the Maximum posteriori (MAP) in the hypobook space [21], that is, the final hypobook which can mostly explain the observed data. Model-based algorithm need to be divided into two independent processes: model training process and prediction process. The former process is usually a process of iterative computing on a training dataset which need a lot of computation resources. Thus, it is appropriate to compute offline in practical application. As the law contained by training dataset is discovered by training process, the calculate cost of online prediction of Model-based algorithm is low. Taking PLSI model as an example, the calculate cost for a prediction is a linear function of the number of hidden classes k.

Another method is to store all the examples simply, the generalization work of these examples are put off to the moment when prediction has to be done. When the learner encounters a new query instance, an objective function value is assigned to the new instance by analyzing the relationship between previous instance and the instance to be rating. This method is called Memory-based or instance-based method [2, 3]. The basic instance-based learning methods in the field of machine learning are k-nearest neighbor algorithm. This algorithm assumes that all instances correspond to the point in n-dimensional space $\mathbb{R}^N$. An instance's nearest neighbor is defined according to the Euclidean distance. For continuous value of the objective function, k-nearest neighbor algorithm takes the weighted sum of the k nearest examples' corresponded value as the predicted results. The biggest difference between Collaborative filtering which adopt instance-based methods with standard machine learning methods is the methods of measuring the distance. Because of the high dimension of instance vector space, as well as the impact of data sparseness, using Euclidean distance to measure the proximity of the node is not usually a good choice. Studies have shown that when the dimension is high enough, the Euclidean distance metric between any nodes would be meaningless [160]. Therefore, people developed several similarities calculating formula to replace the traditional Euclidean distance metric [2, 4, 20, 110, 161, 162]. Compared to modeling methods, instance-based approach has a key advantage that they are not in the instance space to a one-time estimate of the target function, but to make local estimates for each new instance to be predicted. Weighted average of the k-th sample, this method has a good robustness of noise in the training data. The disadvantage of this method is that the online prediction need to search the entire sample space, so the computational overhead. Solution, including the similarity calculation off [4, 5], using the calculation based on the comment on the similarity of the object [5, 24].

From the viewpoint of prediction accuracy, there is no absolute good or bad between instance-based methods and model-based approach. Although a lot of work claimed that compared to the model-based algorithm can get higher prediction accuracy compared with instance-based algorithm, this assessment is entirely empirical. Selection of data sets, the experimental method and the selected indicators will all impact the final result. A comprehensive comparison of prediction accuracy in collaborative filtering algorithm in the early years [2] showed that, instance-based approach in many cases even works better than the algorithm of Bayesian networks, EM clustering and other more complex models. Amazon also uses the instance-based learning methods, and proven to cope with huge traffic [4, 5].

Now back to our question to build the objective function $\rho'(u_x, i_y) : \mathbb{R}^n \times \mathbb{R}^m \to V_I$. Similarly, we can try to use model-based method or instance-based approach to get this objective function. The model-based method faces two main obstacles. First of all, the objective function is optimized for the observed user ratings, and the dependency between the user's trust relationships with the user rating is unknown, the establishment of this dependency model requires additional knowledge. And it is questionable whether this dependency created will ultimately help to find MAP assume. Secondly, when we use a probabilistic model to seek the MAP assume based on the observed user ratings and trust statements, the injected noisy data will have a significant impact on the result, which makes us difficult to use the directed trust networks to resist the attack. Based on these two reasons, in this book we use instance-based learning methods to build a prediction algorithm, it allows us add anisotropy of the trust network, attenuation during the propagation, synbook, etc. to the computation process. The Eq. (6.1) shows the basic principle of this method. By the trust measurement, we select a trusted close neighbor set for target user, and give a corresponding weight for each element in the collection, and then get the predicted ratings through weighted average method. Formally, the trust-based prediction algorithm is very similar to the memory-based collaborative filtering algorithm, the main difference lies in the selection of nearest neighbors and rules for weight calculation. Therefore, the smoothing approach used in collaborative filtering d can also be applied to the trust-based prediction algorithm. We will discuss the impact of different smoothing strategies later.

$$\rho'_{xy} = \frac{\sum\limits_{\{z|\rho_{zy} \neq \phi \wedge u_z \in T_x\}} \rho_{zy} \cdot \omega_{xz}}{\sum\limits_{\{z|\rho_{zy} \neq \phi \wedge u_z \in T_x\}} \omega_{xz}} \tag{6.1}$$

We will design two prediction algorithms, which will use on-normalized and normalized trust measurement respectively. In the non-normalized trust measurement, a new statement made by a node will not affect the importance of his previous statement. While in normalized trust measurement, each node's trust degree to the other nodes is set to a constant proportional to its importance. The more trust statement made by the node, the lower importance of a single statement is. There are many arguments in literature about these two strategies, one viewpoint agrees that only the non-normalized trust metrics can be able to reflect the trust [17, 154], that is the trust of main user x to object z should not increase as the number trust statement made by x increases. Another viewpoint is that the normalized trust metric could reflect the importance of nodes in the trust network better, and can resist the attacks of the malicious nodes [16, 163] at the same time. In this chapter, we will compare the impact of predictive accuracy of these two strategies. In the next chapter, we will conduct a quantitative analysis of the robustness of the algorithm in the two strategies (i.e. the ability to withstand attacks).

6.2 Trust-Based Prediction Methods

6.2.1 Dual Factor Trust Metric

We first analyze the construction of prediction algorithm under the non-normalized trust metrics. Non-normalized metrics is the trust metrics strategy that usually adopted in the literature. The well-known strategies include the Bayesian belief theory [104], TidalTrust [17], MoleTrust [15], etc. In addition, in some studies of collaborative filtering, the user similarity is defined as trust [19], they also use a non-normalized method to derive the indirect similarity among users. In the last chapter, we have introduced MoleTrust and Bayesian belief theory respectively; we also put forward a modified breadth-first search strategy to improve the coverage of the trust metric. Here, we further discuss the problem of such trust transmission in non-normalized metric method.

Considering the simple trust network in Fig. 6.1, we represent the trust degree between different nodes as an integer value in [0, 9]. Using MoleTrust measurement we will get $t_{af} = 8.47$ and $t_{ag} = 9$. This conclusion is clearly inconsistent with the common sense, because all trust degree of d's predecessors is less than f's, and the trust indirectly deduced should not be higher than the predecessors'. TidalTrust is a non-normalization method similar to MoleTrust. It defines the concept of the path intensity: The path intensity of a reachable path of the target node is the minimum value of the arc besides the end arc in the path. TidalTrust uses a reverse recursive method to progressively merge result on the path which has the maximum intensity, and then produce a final result. As the example in Fig. 6.1, TidalTrust measure will get $t_{af} = 8, \quad t_{ag} = 9$, which is still inconsistent

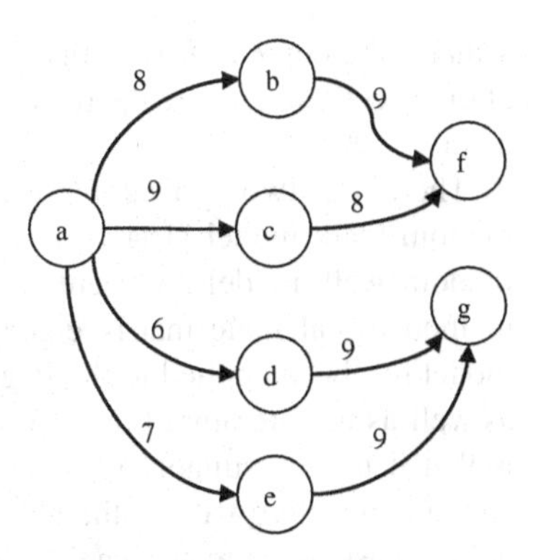

Fig. 6.1 The example of multi-value trust network

with the common sense. This inconsistent phenomenon is called "trust metric amplification".

If we build trust network using Beta trust functions mentioned in the previous chapter, the push operator in belief theory itself can prevent this amplification. But for trust network represented in real value, we should limit their impact on the measurement results of the successors node based on the trust of predecessors. In this book, we improve the weighted formula in MoleTrust (See Appendix 1.1):

$$\omega_{xz} = \frac{\sum_{\{w|u_w \in B_z \wedge \omega_{xw} > TT\}} \frac{\omega_{xw}}{V^t_{\max}} \cdot t_{wz}}{|\{w|u_w \in B_z \wedge \omega_{xw} > TT\}|} \tag{6.2}$$

The transmission of trust also showed attenuation. The research of Golbeck [17] found that, in the case where path intensity is the same, the farther the distance between nodes, the greater the difference of trust given for the same object. In the study of Chap. 5, we also found that, under the condition of same trust metrics results, with the increase of the distance between nodes, the similarity of user opinions decreases. Therefore, when select nearest neighbors in the group prediction algorithm based on trust metrics, the distance among the nodes in the network should also be considered in addition to the trust degree.

We use the following strategy to combine trust degree and node distance into the prediction Eq. (6.1). Based on query (u_x, i_y), trust metrics algorithm first calculates the value of trust of node u_x with other reachable nodes within the specified depth, and record the distance of these nodes. In the prediction process, find the first rated node in accordance with the distance from near and far, and take the depth of the node as the minimum depth, all the nodes whose trust degree is greater than the threshold in the depth compose the collection T_x.

6.2.2 Single Factor Trust Metric

In non-normalized trust metrics method, in order to combine the transitivity attenuation of trust into the prediction algorithm, we represent the results as two scalars: the degree of trust and the distance. Whether can this transitivity attenuation directly reflect the degree of trust? The normalized trust metric provides such a way.

There are two typical representative normalized trust metrics, i.e. Markov random walk model [164], and the propagation activation policy [165]. Markov random walk model represents the trust network in a matrix, and use the power method to calculate matrix eigenvector recursively as trust (or similarity). This model has been applied to the PageRank [164]. EigenTrust [166], SimRank [167], as well as the literature [168, 169]. The limitation of Markov random walk model is that it is only suitable for dealing with non-negative form of trust. For matrix contains negative values, the algorithm may fail to converge, due to the positive definiteness of the matrix can't be guaranteed. The propagation activation policy is more flexible, it can process a negative form of trust. Therefore, we use the propagation activation policy strategy as a normalized confidence measure.

Propagation activation policy has been applied widely inhuman cognitive science [165, 170, 171], as well as information retrieval field [166, 172–175]. Its basic principle is as follows: given a network and initial source points set S, a certain energy e is injected to S and activates these initial nodes. Then, the activated node activates the connected nodes according to certain rules. This process recursively ran until the termination condition is met. The propagation activation can be further divided into unrestricted activation and restricted activation. In the unrestricted activation strategy, each node reached by energy will be activated, and archive some energy. While in restricted activation strategy, in each round of iteration calculation, only the nodes which satisfy certain rules (for example, the energy value is greater than a certain threshold) are able to remain active.

The specific form we adopted for propagation activation policy is similar to Appleseed [16]. Given the source point x and trust network G, first add a virtual trust arc from each node to the source node (set the value as the upper limit of trust), in order to prevent some nodes only gain energy without emission energy. This processing is different from Appleseed, the later takes value coverage strategy. According to our method, more energy will go back to the source point in each iteration, so that the nearer nodes of source node will get more energy when the algorithm terminates. We believe that this set can reflect nearest neighbor principle described in Chap. 3 better. After the modifications to the graph G, the energy e is injected to the source point x. The source point x retained energy $p \cdot e$ $(1 > p > 0)$, and then the remaining energy is split normalized in accordance with the following formula, and sent to all the adjacent points:

$$e_{xk} = \frac{t_{xk} \cdot (1-p) \cdot e_x}{\sum\limits_{u_k \in S_x} |t_{xk}|} \tag{6.3}$$

Each activated node repeats the previous step. In this process, only the nodes whose energy value is greater than 0 can transmit energy. Since absolute value of the sum of nodes' energy in the entire network is a monotonically increasing function, and e is the upper bound of the sum, the convergence of the iterative process can be guaranteed.

It should be noted that, calculating the convergence state of propagation activation process usually takes several times to dozens of times to iterate, it is costly calculated in this way every visitor in the Web environment. Therefore, make restrictions on the number of iterations, which is similar to non-normalized trust metrics. By setting a relatively high p value, we can still give higher energy value to the nodes closer to the source node. All the nodes whose energy value is greater than zero constitutes the trusted neighbor collection T_x in prediction Eq. (6.1), the corresponding energy value of node k in the collection is the weight ω_{xk}.

The improved AppleSeed algorithm can be found in Appendix 2.

6.2.3 User Similarity Weighted

In most applications which allow the user to make trust statement in some form, the manifestation of the degree of trust is usually a single value or two values (trusted and non-trusted). The user can't represent the degree of trust, and the behavior of making trust declaration is in a certain degree of arbitrariness. Figure 6.2 shows the histogram of absolute deviation between adjacent nodes ratings in the trusted network EPP data, the absolute difference is divided in interval of 0–4 and conduct frequency statistics.

From Fig. 6.2, we found that the nodes whose absolute difference are greater than 1.5 also take a considerable proportion, which means that we should not treat the user's trust declaration equally. If the absolute difference of the two user ratings is smaller, the declaration of trust between them should be given a higher weight value. This method can reduce the impact brought by the arbitrariness of user trust statement. Sigma function (Eq. 6.4) is used to deal with weight processing in the single value trust network, where dev_{xy} is the absolute difference defined in Eq. 5.6 in the previous chapter. Under normal circumstances, the user x is not necessarily has common rate items with each user in his trust list, we will set the default value of absolute difference as the average absolute difference of

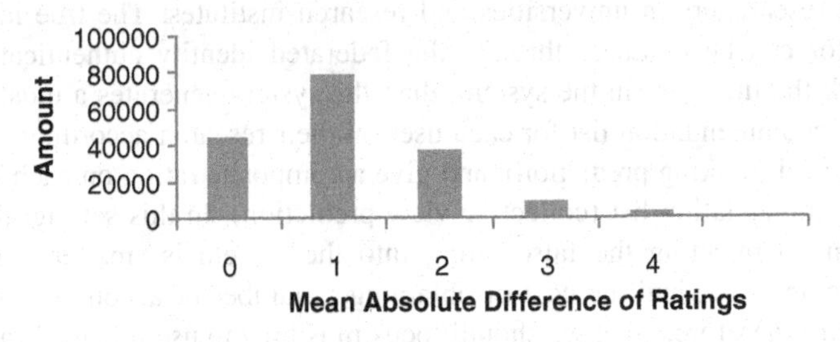

Fig. 6.2 The histogram of absolute difference deviation adjacent nodes ratings in EPP trust network

user x and other users in G_x. If we can't calculate the absolute difference between user x and each node in G_x, set it as the average of global absolute difference. Due to the value range of dev_{xy} is [0, 4], the value range of Sigma function is [0.036, 1].

$$t_{xz} = \frac{2}{1 + e^{dev_{xz}}} \tag{6.4}$$

For binary trust network, we can also use the same way to deal with trust statement and retain a declaration of mistrust (value -1).

6.3 Prediction Algorithm Based on Second-Order Markov Model

In the previous sections, we used the normalized and non-normalized trust measurement method to complete the steps of nearest neighborhood selection and weights calculation, and present one method using similarity to weight the single value or double value trust network, which can decrease the impact caused by the arbitrary statement of user trust. In this section, we will use the idea in the setting of collaborative filtering that utilizing the pass association to complete the steps of nearest neighborhood selection and weights calculation, which can solve the data sparse problem in filtering algorithm.

There is an assumption in collaborative filtering-based algorithms to predict collective views that the user ratings represent their preferences. Using the data of user ratings, we can adopt the memory based learning algorithm or modeling algorithm to capture the similarity between users, and get the prediction result. Because it is difficult to distinguish the authenticity of the rating, the noise data made by the attackers on the Internet have a great impact on the prediction result generated by collaborative filtering algorithms. To defense against such attacks, we propose to utilize the trust network to prevent the attackers manipulating the prediction result of the algorithm by injecting fake ratings with purpose as an auxiliary input.

However, not all the collective view prediction applications are vulnerable to this attack. Considering a literature recommendation system, its target is the students and researchers in universities and research institutes. The true identity of each visitor can be obtained through the federated identity authentication. The users mark the literature in the system, then the system generates a most relevant literature recommendation list for each user on their research according the users' scoring record (ranking prediction), and give a composite rating on each literature in the recommendation list (collective view prediction). In this setting, the users' motivation on injecting the false rating into the system is smaller than in the opening review site, and the cost of obtaining a number of accounts is higher in this system. Therefore, what we should focus in is how to use the user' rating data to generate a ranking prediction and collective view prediction effectively.

Previous research on collaborative filtering frequently used the data set such as MovieLens and Each Movie to verify the algorithm. In this kind of data set, the quantity of rated object is small (from hundreds to thousands), and each user should provide at least a certain number of ratings (for example 20 ratings), this is very different from the typical environment where collaborative filtering applies. For example, the quantity of candidate news recommended on Google News is more than millions every day; the number of review object is hundreds of thousands on Epinion.com. Meanwhile, the number of ratings is smaller, for example, in the EPP dataset, about 62 % of user's rating number is less than 10. In such application environment, we will encounter the data sparseness problem challenges when doing the collective view prediction.

The Fig. 6.3 illustrates the impact of data sparseness problem. Using the user set U and rated object set I as the vertex, the actual rating set E as the arc, we can construct a weighted bipartite graph. The specific weight depends on the range of ratings; it can be a single value representing the visit records or a binary value representing good and bad review or a multi-value representing the degree and so on. As the Fig. 6.3 illustrates, there is no same rating object between the user u1 and user u4 u5, so we can't calculate the similarity between them. In this situation, we can't make a relevance judgment between u1 and u5 or predict the rating of i5 by u1 if we adopt the classic collaborative filtering algorithm.

In some related studies, such as [9, 176, 177], the method that utilizing the pass association rule to solve the problem caused by data sparseness, but these work focus on solving ranking prediction problem, and the output of these algorithms is a value which indicates the extent of relation between the user and object. They are not enough to solve the rating prediction problem. Moreover, when users use multi-value form to indicate rating, two users between which there are many same rating don't always have the high rating similarity. Modeled algorithms, such as PLSI [159] and fuzzy clustering [2], also imply the idea of adopting pass

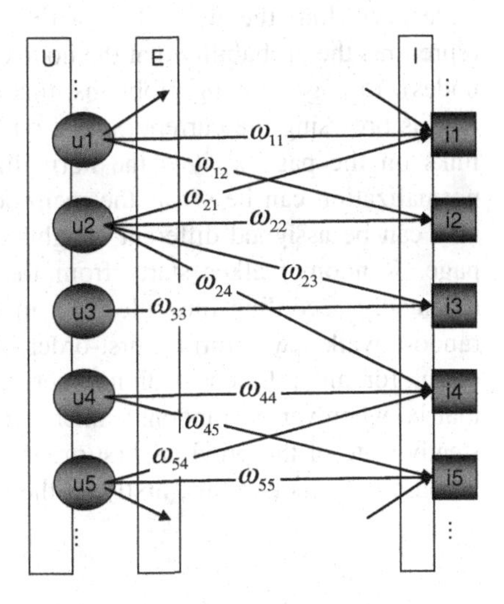

Fig. 6.3 Bipartite graph of Collaborative Filtering system

association. Taking the PLSI for example, it assumes that users and objects belong to a finite number of hidden classes with a certain probability distribution, and uses the Bayesian formula to indicate the probability distribution of observed samples. Then, utilizing EM algorithm, users and objects alternately change their affiliation, until it reaches a local optimal solution. This trained Bayesian model will be used for the prediction. The PLSI algorithm not only can be used for ranking prediction, but also be able to predict the rating with the assumption that the rating that users in each hidden classes marked to the same object obeys a Gaussian distribution. However, this modeled algorithm has a few limitations. Firstly, it assumes that user group can be represented by a finite number of classes, and gets the global optimal solution of the model on the training data, so it can‘t provide the same degree of freedom as the instance-based learning methods. Secondly, in the dataset like EPP, the distribution of user rating is very different from the normal distribution, and this may lead to great prediction errors. Lastly, if a new rating has been add to the dataset, we need to train the model on the global training data again. Because the cost on training is big, model update usually was done offline, so the timeliness of this model is worse than instance-based learning methods.

We propose an instance learning approach based on second-order Markov random walk model. In this approach, the neighbor selection and weight calculating are finished by a second-order Markov random walk process, and then the rating prediction results will be getting by using the weighted average approach.

6.3.1 Random Walk Model

Random walk model was early seen in Page Rank which is a page ranking algorithm. Random walk starts from one or more initial node, then jumps to other node according the defined transition probability matrix (each column vector represents the probability that the node corresponding to this vector jumps to other nodes). In page ranking problems, this corresponds to the probability that a user who is browsing the current page x will go to browse the page y by clicking the links on the page x, it is the normalized out link weights of page x. Weight normalization can be set as the reciprocal of the out-degree of the page simply, also can be assigned different weights according the position of out links in the page. Random walker starts from the initial node, and jumps to other node repeatedly according this rule; in a moment t the probability of the position of random walker will form a first-order Markov chain. According the Ergodic theorem, for an n-dimensional non-periodic and irreducible transition probability matrix, wherever the random walker starts from, after a finite number of jumps, the steady state of the previous first-order Markov process is determined. The calculation process is like this, firstly let the transition probability matrix right multiply any n-dimensional vector whose mode is one, then let the results got before right multiply the transition probability matrix repeatedly, lastly the result will converge to principal eigenvectors of the transition probability matrix. In this way, the page which random walker stops on with a high probability is a higher ranking page. In order to ensure the transition probability matrix is no-periodic and irreducible, the easiest way is to make every jump step to transfer to one or more pre-configured node with a smaller probability. The pre-configured node can be the node which has a higher prior ranking assigned by designer, also can be the set of all nodes (It assume that all nodes have a same importance.).

This random walk model also has been adopted on trust metric (EigenTrust [166]) anomalous detection [168] and similarity metric (SimRank [167]) under the p2p environment. For the past few years, there are many researchers who use multi-dimensional random walk to apply the users' rating and labeling information to ranking prediction [169].

When the user's rating is single-value, regarding the bipartite graph as undirected graph and setting the target user as global transfer node, PageRank algorithm can be apply to similarity measurement between users directly. It subjects to two important properties: 1) if two user nodes point to many same object nodes, they have a high similarity; 2) if these object nodes have a high degree, then their weight in similarity measurement will be lower. This idea is known as TF-IDF in information retrieval. However, if the users' rating is multi-value, the first property may wrong. As the Fig. 6.4 illustrated, we assume that our current task is to measure the similarity between u_2 and other users. There are two same rating objects between u_2 and u_1 and u_3 respectively, and the two objects have a similar degree. Because the rating value of u_2 and u_3 is completely same, the similarity between them should be higher than similarity between u_2 and u_3 as usual. However, if using the first-order Markov random walk model to calculate the similarity, we will get the contrary conclusion. Therefore, when the users rating are multi-value, we couldn't use the first-order transition probability matrix to define the Markov process. In this example, the probability of one user node where u_2 jumps with two steps, not only depends on the arc-value of object node i_2 and i_3 on second jump, but also depends on the arc-value of u_2 on fist jump. If the rating

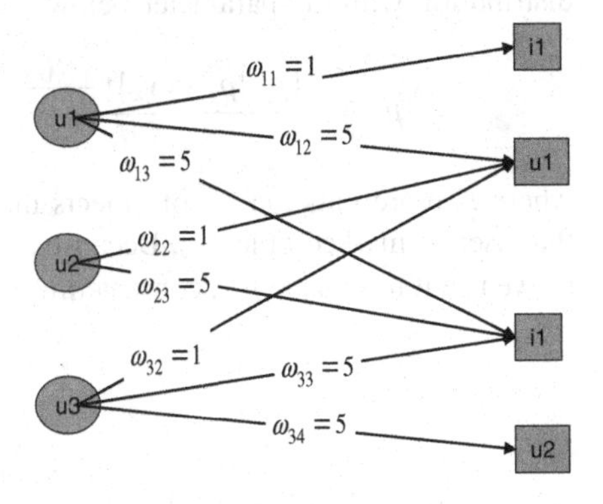

Fig. 6.4 Description of limitations in a first-order Markov process

that someone marks to i_2 and i_3 is closer to u_2, the probability of a two-step jump to this user node is higher. We can imagine this process as that target user looks for the user set in which the user has a similar viewpoint with himself by random walking.

We will adopt the second-order Markov random walk model to solve this problem, the algorithm that adopts this model consists of four steps: (1) Normalizing the users rating to reduce the effect brought by different habit on scoring. (2) Constructing a second-order Markov transition probability matrix; (3) calculating the near neighborhood and corresponding weight by power rate method; (4) generating the rating prediction. Below we will introduce the implement form of those steps respectively.

6.3.2 Algorithm

6.3.2.1 Normalizing the Users Rating

Though the rating can represent the opinion of users on object, different user may have a different view on ratings. Someone is used to providing a high rating, but someone is more cautious on scoring. This difference would mislead us when comparing users' viewpoint on same object. Therefore, normalizing the users rating to compare in the same range is an effect form improving the prediction accuracy.

The usual method used for normalizing in the literatures includes subtracting the average value of all users' rating and utilizing Gaussian smoothing to transfer form, but the two approaches don't take the effect of distribution feature of users rating into consideration. Intuitively, for one rating v_k, the more times the rating which is less than some value appears, the higher extent of user's positive review the rating represents. If the rating v_k appears more times, it represents that a neutral review may go up. Based on the two mutual constraint conditions, we choose an approach called halfway accumulative distribution to normalize the ratings. Given $\rho_{xy} = v_k$, the probability $pref_{xy}$ that user u_x like object y follows the Gaussian distribution with the parameter below:

$$\mu_{xk} = \frac{|\{\rho_{xy}|\rho_{xy} \le v_k\}| - \frac{|\{\rho_{xy}|\rho_{xy}=v_k\}|}{2}}{|I_x|} \quad \delta_{xk} = \frac{|\{\rho_{xy}|\rho_{xy} = v_k\}|}{2|I_x|} \tag{6.5}$$

where I_x represents the set of objects that user u_x rated, ρ_{xy} represents the rating that user u_x marked object i_y. Because u_{xk} represents the expectation of this favor, so we use it to replace ρ_{xy}, constructing a bipartite graph as the Fig. 6.3 illustrated.

6.3.2.2 Second-Order Transition Probability Matrix

We define the probability distribution that target user u_x transfer to the other user node with two jumps as follows. Firstly, u_x jumps to the set of object nodes which he will rating with the probability $\frac{1}{|I_x|}$, this step is same as single-value scoring. The mean of this step is that user u_x treats his rating with the same weight, and hope to look for his neighborhood by checking the users who provide a similar rating. Secondly, on one object $i_y \in I_x$, user checks the rating that each user $u_z \in U_y$ marked the object i_y, then jumps to the node whose rating is close to his with a high probability. The result of the two steps jumping will be calculated by the Eq. (6.6):

$$c_{xz} = \sum_{\{y|i_y \in I_x \cap I_z\}} \frac{1 - |\omega_{xy} - \omega_{zy}|}{\sum_{\{s|u_s \in U_y\}} (1 - |\omega_{xy} - \omega_{sy}|)} \cdot \frac{1}{|I_x|} \tag{6.6}$$

For each user node which user u_x can reach with two jumps will be calculated as the above formula, the results would be expressed in vector form as the x-th column of transition probability matrix. For the node which user u_x can't reach with two jumps, its transition probability would be set to 0.

6.3.2.3 Similarity Calculation

Similarity calculation adopts the standard power rate approach. As the Eq. (6.7) illustrated, in the system consisting of n user groups, $\vec{R}^{(0)}$ is the n-dimension initial vector. Given target user u_x, the x-th element of $\vec{R}^{(0)}$ is set to 1, the other element are set to 0. C is the second-order transition probability matrix. P is the probability of jumping to x, it gives more similarity to the node closing to u_x.

$$\vec{R}^{(t+1)} = (1 - p) \cdot C \cdot \vec{R}^{(t)} + p \cdot \vec{R}^{(0)} \tag{6.7}$$

Similar to the spread activation policies, according the scale of application, above similarity calculation formula not only can stop on steady state ($\|R^t - R^{t-1}\| < \delta$), but also can stop on the assigned times.

6.3.2.4 Predicting Rating

Firstly, we adopt the weight mean Eq. (6.8) to calculate the favor extent of user x to object, in which sim_{xz} is the z-th element of obtained similarity vector R, w_{xy} is the normalized user rating ($pref_{zy}$). Because in the random walk model the user having more rating usual have more chance to obtain trust, we divide a corresponding reversal frequency factor by similarity to balance this effect.

$$pref'_{xy} = \frac{\sum_{\{z|\omega_{zy}\neq\phi\wedge sim_{xz}\neq 0\}} \omega_{zy} \cdot \frac{sim_{xz}}{\sqrt{|I_z|}}}{\sum_{\{z|\omega_{zy}\neq\phi\wedge sim_{xz}\neq 0\}} \frac{sim_{xz}}{\sqrt{|I_z|}}} \tag{6.8}$$

Lastly, we utilize maximum likelihood estimation to transfer the predicted favor extent to rating. The specific practice is shown as the Eq. (6.9).

$$\rho'_{xy} = \arg\max_{v_k \in V} \frac{1}{\sqrt{2\pi}\delta_{xk}} \exp\left(-\frac{(pre'_{xy} - \mu_{xk})^2}{2\delta^2_{xk}}\right) \tag{6.9}$$

6.4 Bayesian Fitting Model

We have given two trust measurement based prediction algorithm and one collaborative filtering algorithm adopting second-order Markov model. A natural question is: can we combine the result generated from different predictor to improve the accuracy of the prediction. In this section, we propose one predictor optimizing approach adopting the Bayesian model. Specifically, given an optional predictor set $\{pd_1, pd_2, pd_3, \cdots, pd_n\}$, because of some unknown features of user, the accuracy of the prediction results generated by these predictors is different for user u_x. We hope to compare the real rating records of user u_x with the outputs of each predictor by Bayesian learning, calculating the probability that rating of user can be predicted by pd_n accurately. Then, we can use the Bayesian formula to calculate the optimal prediction result of user rating.

Firstly, we assume that for the observed rating ρ_{xy} the probability that it was generated from one predictor pd_k is proportional to the Gaussian distribution below:

$$P^k_{xy} \propto \frac{1}{\sqrt{2\pi}\delta} \exp\left(-\frac{(\rho_{xy} - \rho'_{xyk})^2}{2\delta^2}\right) \tag{6.10}$$

For the set of all observed object that user u_x has rated, assuming each rating is mutual independent, the probability that it was generated from one predictor pd_k is like this:

$$P^k_x \propto \prod_{i_y \in I_x} \frac{1}{\sqrt{2\pi}\delta} \exp\left(-\frac{(\rho_{xy} - \rho'_{xyk})^2}{2\delta^2}\right) \tag{6.11}$$

For the convenience of calculation, we assume that the standard deviation of all the predictors is 1, and omit the constant term, getting the following total probability formula:

$$P_x^k \propto \exp\left(-\frac{\sum\limits_{i_y \in U_x} \left(\rho_{xy} - \rho'_{xyk}\right)^2}{2}\right) \tag{6.12}$$

Moreover, the set u_x of rating of user u_x generated from one predictor $\mathrm{pd_k}$ constitute the hypobook space to learn, so the constrain condition is:

$$\sum_{k=1}^{n} P_x^k = 1 \tag{6.13}$$

Synthesizing the Eqs. (6.12) and (6.13), we can obtain the maximum a posterior probability that the set I_x of all ratings of u_x was generated from predictor $\mathrm{pd_k}$. Obtaining the parameter of the Bayesian model by training it, for new unknown rating ρ'_{xy}, we can adopt the Bayesian formula to weight the outputs of all predictor to calculate the ρ'_{xy}:

$$\rho'_{xy} = \sum_{k=1}^{n} \rho'_{xyk} \cdot P_x^k \tag{6.14}$$

6.5 Experiments

In this section, we will do a horizontal comparison for the prediction accuracy and prediction coverage of trust-based prediction algorithm mentioned above, the similarity-weighted prediction algorithm, and the second-order Markov random walk prediction algorithm. Leave-one-out and k-folders crossing validation are the standard form of measuring the accuracy of learning algorithms. Leave-one-out takes out one rating from the given set D of ratings of users each time, and uses the rest of ratings and the other available information to predict, then repeats this process for all ratings of set D, and uses the selected performance metric to analyze the results. In the k-folders crossing validation, the ratings of each user are divided into k parts equally, one part is put into testing dataset, the others are put into training dataset. Repeating k times, we will get k orthogonal training dataset and testing dataset. Training dataset is used to generate prediction; testing dataset is used for results comparison. In the leave-one-out, after getting a prediction, we need to train the algorithm anew. However, in the k-folders crossing validation, calculation results of one training or neighborhood selecting can be used to generate more than one prediction results.

Considering the calculation cost, we adopt the k-folders crossing validation and its variants to do the evaluation. The specific scheme will be introduced in the following section.

6.5.1 Evaluation Measurement

We adopt two prediction accuracy metrics to compare the performance of these algorithms which was used widely in literature [2, 16], including mean absolute error (MAE) and prediction coverage.

MAE compares the predicted rating with user's real rating, and takes the arithmetic average for all the results. Specific definition is as follows:

$$\mathrm{MAE} = \frac{\sum_{\rho'_{xy} \in PT} |\rho_{xy} - \rho'_{xy}|}{|PT : \{\rho'_{xy} | \rho_{xy} \in RT \wedge \rho'_{xy} \neq \phi\}|} \tag{6.15}$$

Which, RT represents testing dataset. Because of the effect of sparse user's ratings, not a prediction algorithm under any circumstances is able to produce the predicted results. We use PT to represent the set of predicted rating that a prediction algorithm is able to generate. Lower MAE means higher prediction accuracy.

Prediction coverage is defined as the ratio of the rating that can be predicted and all ratings in the testing dataset. Higher prediction coverage means higher prediction performance.

$$\mathrm{Coverage} = \frac{|PT|}{|RT|} \tag{6.16}$$

6.5.2 Trust-Based Algorithms

We evaluate the MAE and prediction coverage of trust-base prediction algorithm with basic prediction formula and weighted similarity. For comparison, we use Pearson correlation coefficient (formula 6.1) and Cosine distance (formula 6.3) to measure the similarity between users, and adopt Eq. (6.1) for prediction. In the following horizontal comparison of various algorithms, we will evaluate the effect on results which user ratings normalization has.

We adopt the EPP dataset as the experiment dataset, whose statistics distribution can be seen in Chap. 5. We don't adopt the dataset of DianPing.com for evaluation, as mentioned before, due to the restrictions of user privacy protection policy, the friends list of some nodes can't be collected in this dataset, which would adversely affect the objectivity of the assessment results. The EPP dataset doesn't provide the rating that user rerated on review object, so it isn't suitable for algorithm assessment.

In the MAE evaluation, we find out all users who provide at least one trust statement and two ratings firstly, then we extract one rating from rating records of each user, constructing the testing dataset, and calculate the prediction errors of different algorithms. Repeating above back into pack sampling process ten times,

Table 6.1 Comparison of prediction accuracy and prediction coverage between different algorithms

	PCC	Cosine	Dual factor trust	Single factor trust
MAE	0.863	0.867	0.892	0.893
Prediction coverage (%)	71.0	72.2	85.3	

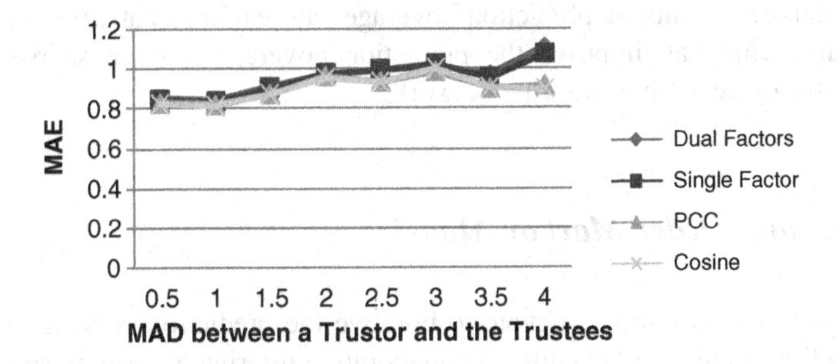

Fig. 6.5 The relation between prediction accuracy and the absolute difference between rating of user and rating of other users in the trust list

we take the mean value of results of ten times lastly. By this way, we give all users same weight in the assessment results.

The results of different algorithms are shown in Table 6.1. The prediction result of single factor trust algorithm is obtained in the case that the parameter p is set to 0.5. From this table, we can find that the prediction accuracy of two trust-base algorithm and two typical collaborative filtering algorithms is close. On the whole, the prediction result of collaborative filtering algorithm is better. Similar conclusion also can be found in the research of Massa. This seems to indicate that it doesn't lead to better prediction accuracy that simply taking the trust network as the basis of neighborhood selecting and weight calculating in prediction algorithms. In order to explain the reason for this phenomenon, when calculating the MAE of each user, we also record the mean absolute deviation between his rating and the rating of the users in trust list. Sorting the mean absolute deviations from low to high and calculating the mean of MAEs with a 0.5 interval, we get the results shown in Fig. 6.5. From the results, we can find that the absolute deviation of rating between user and the other users in his trust list has a great effect on the accuracy of prediction algorithm. In the interval [0, 2.0], the prediction accuracy of these algorithms is very close. With the absolute deviation of rating increasing, the prediction accuracy of the trust-based algorithm keeps decreasing, but the collaborative filtering algorithm is stable. This phenomenon explains that users can understand the mean of trust statement correctly, and it is the key factor affecting the accuracy of prediction algorithms that users manage their trust list according this. For those cautious users, the trust-based prediction algorithm can obtain a better result.

It should be noted that we don't have the user rating normalized in this experiment. Different user has a different understanding on rating, which will affect the prediction accuracy. For example, user x may tend to provide a low rating, but the comment he writes has a high quality. After seeing the comment of user x, user z will add user x to his trust list, but he tend to provide a higher rating. In this situation, normalizing the user ratings contributes to improve the accuracy of prediction algorithm. We will discuss further the solution for this problem later.

The statistics results of prediction coverage rate explain that trust-based prediction algorithm can improve the prediction coverage rate on sparse dataset prominently by searching the trust network.

6.5.3 Second-Order Markov Model

Second-order Markov model devote to improve the prediction coverage rate and the prediction accuracy of traditional collaborative filtering algorithm, so we need to evaluate the performance of this algorithm when facing different sparse extent of data. Therefore, we adopt one variation of the k-folders crossing validation to complete the experiment. Firstly, we put a certain number of user ratings into training dataset, which stand for the mature users who provide much more ratings in the system. Secondly, we put 5, 10, 15, 20 ratings of each test user into training dataset respectively, and then use the rest of his ratings to test the prediction accuracy and coverage. By this way, we can obtain some dataset with different sparse extent. Accordingly, we name these datasets for given 5, given 10, given 15 and given 20 respectively.

We adopt two different datasets in this experiment, which come from EPP dataset and MovieLens dataset respectively. According to the need of the testing approach, we only keep the data of the users, who have at least 40 ratings in the two dataset. The preprocessing statistics is shown in Table 6.2, from this table, we can see that the sparse extent of EPP dataset is much more than MovieLens dataset. As a comparison, we evaluate the collaborative filtering algorithm with PCC and cosine distance as its similarity calculation approach on the same dataset. Because we normalized the user ratings in the second-order Markov model, to PCC and cosine prediction algorithms, we use prediction Eq. (6.2) to normalize the

Table 6.2 Dataset summary in the experiment

Attribute	Epinion (EPP)	MovieLens (ML)
Number of users	3760	645
Number of objects	106742	1681
Number of ratings per user	92.5	142
Sparse extent (%)	99.9	91.5
Possible ratings can taken	{1, 2, 3, 4, 5}	{1, 2, 3, 4, 5}

Table 6.3 The MAE results of different prediction algorithms in EPP dataset

Dataset	Algorithm	Given5	Given10	Given15	Given20
EPP200	PCC	1.000	0.967	0.938	0.912
	Cosine	0.999	0.958	0.927	0.903
	SORW	0.969	0.903	0.866	0.843
EPP400	PCC	0.998	0.952	0.924	0.902
	Cosine	0.955	0.946	0.915	0.891
	SORW	0.960	0.892	0.856	0.836
EPP600	PCC	0.967	0.928	0.903	0.887
	Cosine	0.961	0.917	0.892	0.874
	SORW	0.942	0.876	0.843	0.828

Table 6.4 The MAE results of different prediction algorithms in ML dataset

Dataset	Algorithm	Given5	Given10	Given15	Give20
ML100	PCC	0.855	0.819	0.798	0.783
	Cosine	0.826	0.778	0.764	0.756
	SORW	0.838	0.778	0.762	0.751
ML200	PCC	0.837	0.803	0.787	0.772
	Cosine	0.817	0.774	0.757	0.747
	SORW	0.833	0.775	0.749	0.735
ML300	PCC	0.840	0.800	0.783	0.775
	Cosine	0.814	0.767	0.748	0.741
	SORW	0.847	0.773	0.747	0.739

user ratings, namely each rating subtracts the mean rating of this user, to compare the prediction accuracy of different algorithm objectively.

Because the EPP dataset has more users, we randomly put the rating data of 200, 400 and 600 users into training dataset respectively. In the experiment of MovieLens dataset, the number is 100–300. The results of MAE are shown in Tables 6.3 and 6.4 respectively. We can see that, on sparse EPP dataset, the prediction accuracy of second-order Markov model is better than PCC and Cosine algorithm. On ML dataset, several algorithms have close prediction accuracy. The results in the tables are obtained in the condition that the parameter p is 0.5 and the maximum depth is 3.

The comparison of prediction coverage rate is illustrated as the Fig. 6.6. We can see that in the EPP dataset the prediction coverage of second-order Markov model is much better than the PCC and the Cosine similarity algorithm. When user can provide more than 10 ratings, the prediction coverage rate is close to 100 %. On the ML dataset, because of the low sparsity, the prediction coverage rate of all prediction algorithms is close to 100 %.

In the Fig. 6.7, we show further that how parameter p affects the prediction accuracy of the second-order Markov model. The curve in this Figure is obtained by averaging the results in the Tables 6.3 and 6.4 with a different parameter value. We can see that the parameter p don't have a great effect on the results.

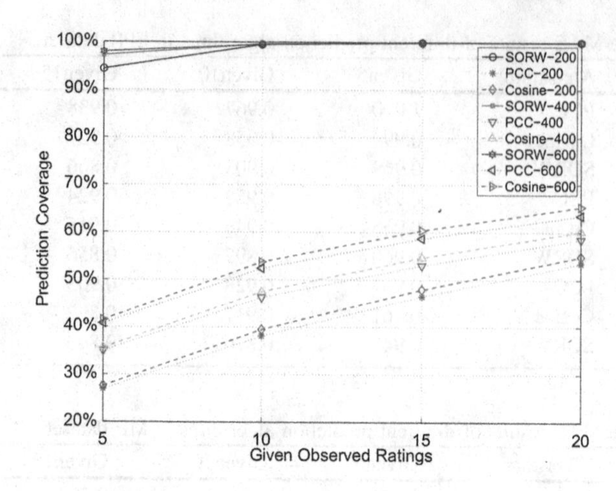

Fig. 6.6 Comparison of prediction coverage of different algorithms in EPP dataset

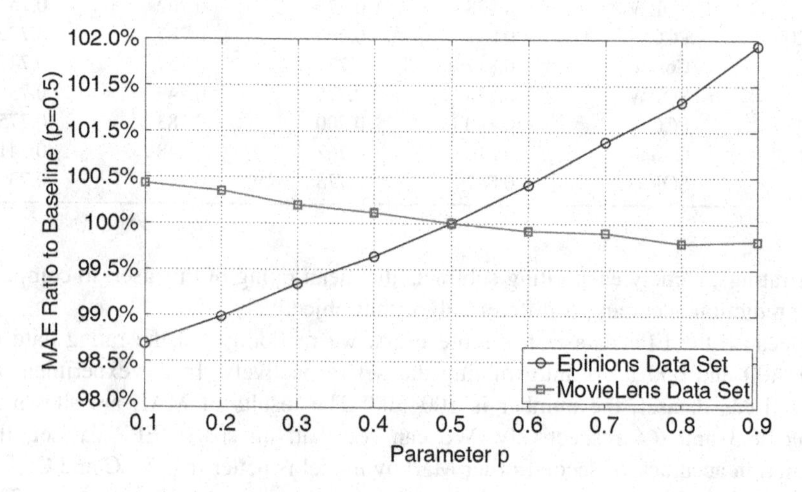

Fig. 6.7 The impact of parameter p on prediction accuracy

6.5.4 Bayesian Fitting Model

We take the data of the user whose average rating is more than 25 from the EPP dataset to construct the dataset of this experiment. We will put 3 in 5 rating data into training dataset A1 into training dataset B1 into testing dataset. Utilizing the training dataset B as the observed variable and training dataset A to yield the prediction result, we adopt the Bayesian model above to train a fitting model for each user, and then use the testing dataset and training dataset A for testing, to confirm whether Eq. (6.14) can improve the prediction accuracy. In the process of testing, the training dataset and the testing dataset should keep orthogonal.

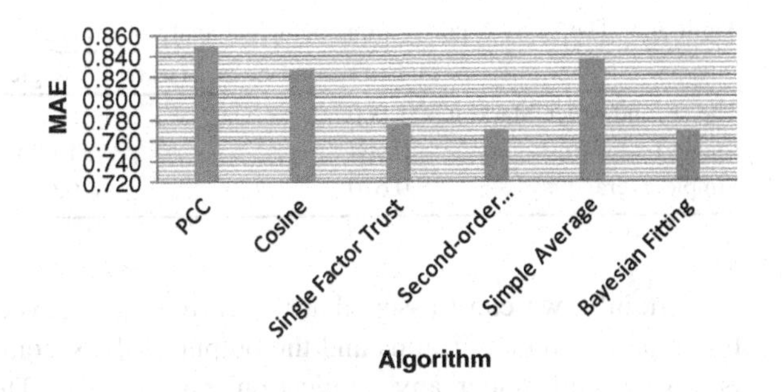

Fig. 6.8 The comparison of *MAE* among different algorithms

In the experiment in this book, we adopt the outputs of three predictors to train the Bayesian model above, which is the trust-based prediction algorithm, the two-second Markov random walk model and simple average method. We don't adopt the PCC and Cosine collaborative filtering algorithm, because the prediction accuracy of these two algorithms is lower, and not fit to train the Bayesian fitting model. Moreover, according the evaluating result above, the prediction accuracy of these two algorithms is lower than the second-order Markov random walk model. On the selection of the trust metric-based prediction algorithms, because the prediction accuracy of single factor and double factor prediction algorithm is close, so we only adopt the single factor prediction algorithm to construct the Bayesian fitting model, and adopt the same user ratings normalization policy with the second-order Markov model.

The experimental result is illustrated as the Fig. 6.8. From this figure, we can see that the prediction accuracies of the Bayesian fitting model, single factor trust prediction algorithm and the second-order Markov random walk model are very close, and all of them are better than traditional collaborative filtering algorithm and the simple average algorithm. Comparing the results in Table 6.1, we can find out normalizing the user ratings can improve the prediction accuracy of various algorithms.

The experimental results also show the improvement in prediction accuracy taken by the Bayesian fitting model is very little on the dataset we adopted. In order to identify the reason, we analyze the relation among the trust-based prediction algorithm, the second-order Markov model and simple average algorithm. Table 6.5 shows that the Pearson correlation coefficient between the single factor trust-based prediction algorithm and the second-order Markov model is 0.964, which shows these two algorithms can be replaced each other. However, the prediction error of the simple average algorithm is big, leading to that the fitting results of the Bayesian model tend to give these two algorithms before a bigger probability value. The comprehensive results of the above factors make the performance improvement taken by the results of the Bayesian fitting model very little. This proves the conclusion in the Chap. 3 again from the opposite angle, that the trust network contains the correlation among user views. And the prediction algorithm proposed in this book can capture this correlation better.

Table 6.5 The Pearson correlation coefficient analysis among the outputs of the three predictors

	The second-order Markov	Single factor trust	Simple average
The second-order Markov	1.000		
Single factor trust	0.964	1.000	
Simple average	0.670	0.661	1.000

Certainly, we can't assert that the correlation between the output of the trust-based prediction algorithms and the output of the second-order Markov algorithm is always high under any application environment. Theoretically, the Bayesian fitting model can select a fit predictor for each user to generate a better prediction result.

6.5.5 Complexity Analysis of Algorithm

The algorithm complexity of the trust-based prediction algorithm and the second-order Markov model-based algorithm consists of two parts: (1) calculating the neighborhood set and corresponding weight; (2) utilizing the weighting formula to generate prediction result.

For steps one, the algorithm complexity of these two adopted trust-based algorithms above depend on the average out degrees $\bar{d}$ of each node in the trust network G and the search depth max dep. The algorithm complexity of these two algorithms all is $O\left(\bar{d}^{maxdep}\right)$, it increases exponentially with the average out degree of node and the maximum search depth increasing. For the second-order Markov model-based algorithm, its algorithm complexity also depends on the average out degrees and search depth of each node in the corresponding graph of the second-order transition probability matrix, which is also $O\left(\bar{d}^{maxdep}\right)$. Therefore, when the search depth of these two algorithms is the same, the algorithm complexity mainly depends on the average out degrees of each node. In the pre-processed EPP dataset adopted in the Sect. 6.5.4, the average out degrees of each node in the trust network is 33.2, but the average out degrees of each node in the corresponding graph of the second-order transition probability matrix is 867.2. Obviously, the efficiency of the trust-based prediction algorithm is better than the second-order Markov model-based algorithm. We think the latter should be adopted on the sparse dataset, or the algorithm complexity isn't satisfied on very large dataset. Or we only adopt the second-order Markov model-based algorithm on the user who provides a little of rating in dataset, in order to solve the "cold start" problem.

The algorithm complexity of utilizing weighting formula to yield the prediction result depends on the average number of rating on target object $\bar{r}$, and the data storage structure of the result in step one. If the result is stored in the dictionary form, prediction only needs to traverse the ratings on the target object, the algorithm complexity is $O(\bar{r})$.

6.6 Summary

It makes us be able to utilize neighborhood learning theory to construct a reasonable collective view prediction algorithm that the correlation between the distance among nodes in the trust network and the similarity between trust degree and collective view. In this chapter, we studied the adaptability of two kind of trust measuring method to the prediction algorithm based on the neighborhood learning. The experiment shows that different trust measuring policy has a little effect on the prediction accuracy. Moreover, normalizing the user ratings can improve the prediction accuracy prominently.

We also expanded the traditional collaborative filtering algorithm, and studied how to solve the data sparse problem by the second-order Markov random walk model. The experiment shows that the prediction accuracy and prediction coverage rate of the new approach is better than the typical PCC and Cosine collaborative filtering algorithm. The experimental result shows that the correlation degree between the output of the second-order Markov model-based algorithm and the trust-based algorithm is very high on the Epinions.com dataset, so the improvement on the results by Bayesian fitting model is not significant

Overall, the trust-based collective view prediction algorithm proposed in this chapter can provide close prediction accuracy with the collaborative filtering algorithm with the same principle. All of them are better than the typical collaborative filtering algorithm. Comparing them, we can find out that the algorithm complexity of the trust-based prediction algorithm is lower.

So far, we have proved the trust relation data of the user can improve the accuracy of the collective view prediction. However, can the trust measurement-based prediction algorithm resist the attack of network enemies? In the next chapter, we will propose a formalized analysis framework for this problem, and evaluate the two kinds of trust-based algorithms proposed in this chapter.

Chapter 7
Framework for Robustness Analysis

The authenticity of comments (ratings) and feedback from users have a direct impact on view prediction in the collective system. Authenticity refers to the user's comments and feedback which represents their subjective opinions. For example, if user *x* is satisfied with product *y*, he will comment a high rating for it, and if user *x* give user *z* a trust statement that means user *x* believe that user *z*'s comment is helpful. If most users in the system cannot provide correctly information (comments and trust tags), it's hard to estimate the ratings which the target users will give to unknown item effectively in collaborative filtering and trust-based prediction system. There is an extreme bad situation: every user in system chooses and ratings item randomly. In this situation, any prediction method is meaningless because the items' rating is uniform distribution. Fortunately, this circumstance is nearly impossible in normal user groups. Most users understand the meaning of the rating and feedback, and provide the information faithfully to improve the quality of the service. That's why collaborative filtering has been successfully used in Amazon.com, Google news and Netfilx.com.

However, since people increasingly notice the huge influence of collaborative view in social network, some hackers have plenty incentive to cheat in the collaborative prediction algorithm for economic benefits. As Sect. 6.1 shows, attacker can make false rating to significantly affect the output of collaborative filtering algorithm. In [6, 7, 11] this behavior which inject noise data has been named "attack". Even though trust metric can resist such simple attack, attackers also can find weakness to attack when they understand system's mechanism. Therefore, it's necessary to evaluate systematically the ability of trust-based prediction algorithm against attacks. In this chapter, we will give formal definition for trust-based prediction algorithm attack, attack power and algorithm robustness, and we will analysis common attack strategies and the trust-based prediction algorithm robustness. In the end, we find out some weak points in the algorithm through qualitative analysis and simulation experiment, and make some improvement strategies.

The assessment model also provides general methodology guidance for evaluating other trust-based view prediction algorithm.

T. Luo et al., *Trust-Based Collective View Prediction*,
DOI: 10.1007/978-1-4614-7202-5_7, © Springer Science+Business Media New York 2013

7.1 Related Work

O'Mahony and his partner are the first pioneers to research robustness about Collective view prediction. They focus on analyzing the change of output of collaborative filtering algorithm in the face of attacker deliberately making noise data. It's much different from others who put their attention into accuracy of the algorithm. They made the change degree of output as main measure of robustness. The main contribution of them is their formal definition about attack power, algorithm robustness. They also analyzed PCC and Cosine as metric for collaborative filtering algorithm robustness. The results show that attacker can deliberately make noise data in rating matrix for affecting the output of algorithm. In study [11], author proposed much more attack strategies and broader the framework of algorithm, they were also referenced in [10–14].

However, O'Mahony's framework cannot be directly applied in evaluating trust-based prediction algorithm. First, attacker cannot simply add false data to influence the output, he need to conquer some point in the trust network to add false rating or change the architecture of network to affect the input. What is possible attack strategy to trust network? How much the impact of the attack? The solution lay on new analysis framework. Secondly, O'Mahony analysis framework only focus on the instant robustness of prediction algorithm, while in actual environment the input of algorithm is dynamic. For example, when the false comments and ratings are discovered by some users, these users can give rating to the false comments or modify the trust tags of attacker. It's beneficial for resisting attack, and the dynamic assessment ability is also what we expect.

In this chapter, we proposed an assessment framework of trust-based collective view prediction algorithm to meet the requirements. The framework inherits some basic concepts from O'Mahony such as attack power, robustness, and attack strategies. In addition, we assume trust network G is created by user trust statement, but it can also be applied to trust function based trust network.

7.2 Definitions

Collective view prediction is dependent on item ratings from community members. According to O 'Mahony' definition, in the collaborative filtering system, given the user-item rating matrix D, the attack T_D is a transformation from D to D', and make D' different from D. The attacker can complete the conversion process by creating a certain number of accounts, and adding, deleting or modifying the records in rating matrix.

In trust-based collective view prediction, neighbor-choose and weight-value depend on trust network, so simple attack does not significantly influence the output results of the algorithm. In addition, the attackers' behavior of creating some accounts to add trust-tag into the network also cannot affect the output of the algorithm, because there is no arc toward the false node. Enemy must control some nodes and make these nodes connected to the false node.

Definition 1 Given the user rating matrix D and trust network G, attack T_D is a transformation from D to D', and where D' is a different matrix from D. Attack T_G is a transformation from G to G', where some of nodes in G' are controlled by the attacker.

In this definition, we described the behavior of the attack to trust network as nodes change in G. For example, the attackers first create some accounts, and use them to publish some genuine comments. After a period, these accounts will obtain some in-degree. The attacker also could bribe community members to come to the same purpose. In trust-based prediction algorithm, T_G and T_D must occur at the same time to affect the output results.

In order to influence the output of algorithm, the attacker must consume some resources to complete T_G and T_D. We set all possible type T_G to be γ_G, type T_D to be γ_D, and define cost function $C_G : \gamma_G \to R_G$ as the cost resources to attack trust network and define cost function $C_D : \gamma_D \to R_D$ as the cost resources to attack the users' rating matrix. The detail of cost function depends on application. For example, C_D is the monotony increasing function to add ratings into D, it reflects the attacker's cost of time and effort to create such false comments. High quality comments usually tend to attract users to make trust statement therefore C_G is defined as the monotony increasing function to create high quality comments.

When the attacker gets the accounts through T_G and modifies the set D, the output of the prediction algorithm may be affected, and we call the influence degree as attack strength, its specific defined as follows:

Definition 2 Given attack T_D and T_G, the attack strength to item i_y is

$$POA(i_y, D, G, T_D, T_G) = \frac{\sum_{\{u_x \in U | \rho_{xy}, \rho'_{xy}, \rho''_{xy} \neq \phi\}} \frac{|\rho'_{xy} - \rho''_{xy}|}{Max(|R_{\max} - \rho'_{xy}|, |R_{\min} - \rho'_{xy}|)}}{|\{u_x \in U | \rho_{xy}, \rho'_{xy}, \rho''_{xy} \neq \phi\}|} \tag{7.1}$$

The cost of T_D is c_D, and the cost of T_G is c_G. ρ'_{xy} is the output before attack and ρ''_{xy} is the output after attack. In the above definition, we measure the attack strength by the change of the output of prediction algorithm in D, because community members' attention is different to various comments. For example, the user x may only pay attention to electronic products, and the user z focused only on cosmetics. When the attack is only aimed to electronic products, there is no affect to user z. The ratings in D are the direct reflection of such diversity, so it's an appropriate metric to measure the influence of attack. So we define the robustness of the system as follows:

Definition 3 Given the attack T_D and T_G, the robustness of the system is

$$robustness(D, G, T_D, T_G) = 1 - \frac{\sum_{i_y \in I} POA(i_y, T_D, T_G) \cdot |\{u_x \in U | \rho_{xy}, \rho'_{xy}, \rho''_{xy} \neq \phi\}|}{\sum_{i_y \in I} |\{u_x \in U | \rho_{xy}, \rho'_{xy}, \rho''_{xy} \neq \phi\}|} \quad (7.2)$$

where I is the comments set.

We give higher weight to the object which has more rating, as it represents how much attention people have paid to it. The ρ'_{xy} and ρ''_{xy} in Eq. (7.2) were calculating by leave-one-out method which is a standard method in machine learning. This method takes off a rating ρ_{xy}, and predict ρ'_{xy} by the remaining D and G, or predict ρ''_{xy} by the data from T_D and T_G.

7.3 Noisy Data Injection Strategy

The attacker needs to complete the following steps to infuse noise data for affecting the output of prediction algorithm. First, it need to control the in-degree nodes set U_T in G. Otherwise in trust-based prediction algorithms, the compromised nodes cannot have an impact on the others. Second, the attacker could directly use the occupied to rating object (by writing comments) and change the output of prediction algorithm, we name this attack as "direct attack". The attacker also could create some new accounts U'_T, and build the trust statement from U_T to U'_T, then use these new accounts to write comments for item. In addition, the attacker could change the trust statement from U_T to $U - U_T$, it also could influence the output, and we named it as "Sybil attack".

The two types of attack as shown in Fig. 7.1, T_G represents controlling the node set U_T which have in-degree in G; T'_G represents creating new accounts set U'_T and building or modifying the trust statements from U_T to U'_T; T_D represents the attacker make false rating to modify D.

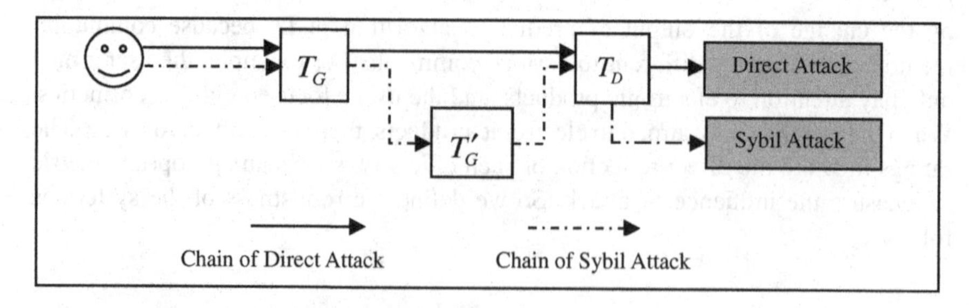

Fig. 7.1 The steps of direct attacks and Sybil attacks

The attacker has the following two intentions for attack D. The first is improving comments for specific item to attract people to buy the item, which is named "pushing attack" and represented with T_D^p; The second is reducing the prediction rating to smear competition, which is named "distorted attack" and we use T_D^n to present it. In the "pushing attack", the attackers usually mark maximum ratings for specific item if system permit, while "nuking attack" is inverse. As users usually marks high ratings for some items in an actual system, so the nuking attacks usually make larger effects than pushing attacks with the same attack cost and attack strategy.

7.4 Attack Cost

Each kind of noise data is corresponding to a cost when the attacker injected strategies, we use c_G, c'_G and c_D to represent them.

The attack cost c_G depends on application environment. Usually we suppose it is linear correlated: U_T is controlled accounts set, *Ind* is in-degree nodes set where $c_G \propto \sum_{u_x \in U_T} Ind(u_x)$. In this assumption, the cost and the total in-degree accounts in G is proportional. In a given attack cost, we consider two attack strategies: random attack T_G^r and concentrated attack T_G^f. In random attack, the attackers randomly choose nodes from non-zero in-degree accounts in G till the cost is equal or greater than c_G. If the *n*th choice result makes the cost more than c_G, the attacker tries to choose another node randomly in order to exhaust c_G in concentrated attack. We assume that the attacker chooses nodes according to the decreasing order in in-degree, till the cost is equal or more than c_G. If the cost is more than c_G in the *n*th node, then the attacker tries to choose another node.

Attack cost c'_G is irrelevant statistic with G, it depends on the false accounts U'_T which are created by controlled accounts U_T, and the trust statement from U_T to U'_T. In theory, the attackers can create a large U'_T, and make all nodes which trust U_T indirectly trust U'_T. In this study, we will analysis attack strength for c'_G.

The attack cost c_D depends on the cost to create false ratings. In order to evaluate various attack strategies for trust network G, we assume that attackers make full use of U_T and U'_T. in every attack T_D, and c_D is in proportion to accounts which make attack. This assumption is valid in Epnions.com and public comments web, because these system permit every account publish one comment in a certain time for one item.

7.5 Feedback Effect

When the attack T_D happened, the honest users in the system can modify the trust statement on false comments to report such behavior. In the experiments followed, we assume that the false comment on specific item i_y can be discovered by the normal users U_y who have already written a comment on i_y, and the authors of the false comments will be joined into their distrust list. This assumption just reflects a probability distribution of normal users' distrust to the authors of false comments. If one item has many comments, it means that many users pay attention to it, so the probability of normal user making distrust statement for the attacker is corresponding high.

The users' feedback is different in reducing the attack strength for direct attack and Sybil attack. In direct attack mode, normal users U_G will modify the trust relationship with nodes in U_T. This local behavior will affect other nodes' trust metric because of the transmitting in the trust network, thus gradually reduces the U_T node ratings' influence to prediction results. In Sybil attack mode, normal users U_G can observe the behavior of the nodes in U'_T, and modify the trust relationship. Such feedback wouldn't change the trust metric between U_G and U_T significantly. If the attacker is smart, he can build different U'_T in every attack, which will make F_G invalid for improving algorithm robust.

It is very hard to prevent the attack only from algorithm level, so we advise to set a minimum values to reduce attack harm. The method is similar to literature [125], the designer can set the rule: user must publish no less than k comments and every comment needs to exist c days, so his comments and ratings can hold a certain weight value in prediction results. By this way, even attacker create a large set U'_T, the U'_T cannot direct influence the results. The attacker must use the nodes in U'_T to publish enough comments and wait for a certain period. The attacker also needs to write comments with high cost in order to the nodes in U'_T received distrust statement from normal users. In this constraint, the attack cost c'_G presents constraint condition to create a certain number "medium quality" comments.

7.6 Application Assessment

We adopt EPP data (see Sect. 5.3) set to test the robustness of algorithms in typical attack scenarios. Each attack scene follows an attack link in Fig. 7.1. In all scenes we set the sum c_G of top 5 nodes' in-degree in trust network G as the basic cost, and test the influences of random attack T^r_G and concentrated attack T^f_G to results, we also remove rating data of users set U_T from rating data D. In addition, in scene 2 and 4 which tested the dynamic change robustness of algorithm with honest user feedback, we remove false ratings after every attack and then start next step. Through these setting, we can evaluate the robustness of algorithm in the same test data.

In the evaluation, we adopt the double elements prediction (presents as MT) in the Chap. 6 and single element prediction algorithm (presents as AS) and their simple form [see Eq. (7.1)]. Two algorithms maximum trust metric set as 3, attenuation factor p set as 0.5 in AS.

In all scenes, we only test the attack strength of "distorted attack", which use controlled accounts to write the lowest ratings permitted by system to attack target i_y in attack T_D. Usually the users tend to offer higher ratings, so the "distorted attack" is easier to change the predication results than "pushing attack" in the same attack strength.

Scene 1: Attack T_G. In this scene, we don't consider the dynamic influence from user's feedback, and assume that the attacker wouldn't start attack T_G. The attack cost c_G set as 1–5 times of basic cost, the results as Fig. 7.2. We can find that the robustness of trust-based prediction algorithm is the same in randomly attack mode (MTRA and ASRA) and concentrated attack mode (MTFA and ASFA), they is only influenced by attack cost c_G. Simple average prediction algorithm presents better than trust-based algorithm in emphatically attack mode (AVGFA), but worse than the latter in random attack mode. In fact, simple average prediction only is influenced by the number of false ratings. The attacker could create large number of false accounts to mark the expected points in target item for attacking. In trust-based prediction algorithm, the strength of attack constrains by the sum of in-degree in U_T.

Scene 2: Attack T_G and F_G. We set c_G as basic cost to test the tendency of robustness with users' feedback. In test process, firstly we randomly choose 1,000 items to build a ordered set $<i_1, i_2, \ldots, i_{1000}>$, then attack every object in sequence. We assume that every user who wrote ratings for the attacked item will make distrust statement to attacker. For simple calculation, we assume that attacker launched 200 attacks in the same time. After 200 attack we observe the change of the dynamic attack strength. From Fig. 7.3 we can see that only 0.72 % items in set I, the trust-based algorithm can recover from attack, and AS algorithm recovery is slow in randomly attack mode.

Scene 3: Attack T'_G. We set c_G as the basic cost, and set each account in U_T can create 50–250 false accounts, the interval is 50. We consider the worst situation: the attacker could easily create trust statement from U_T to U'_T, attack T_D begin from U'_T. Through analysis of this scene, we expect to verify that the handling of normalized and the non-normalized to trust metric has some influence to

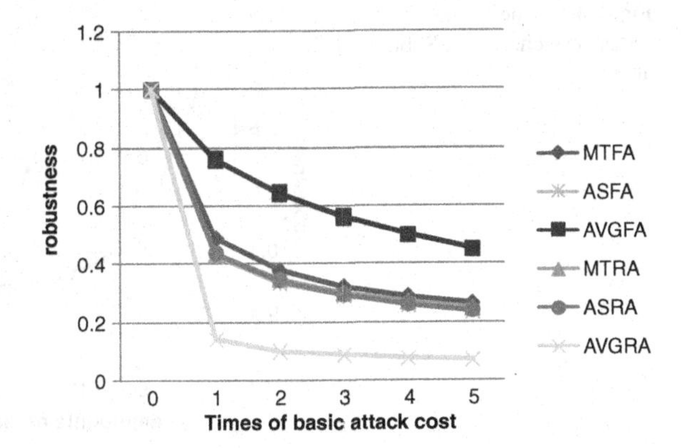

Fig. 7.2 Scene 1: the robustness change in direct attack

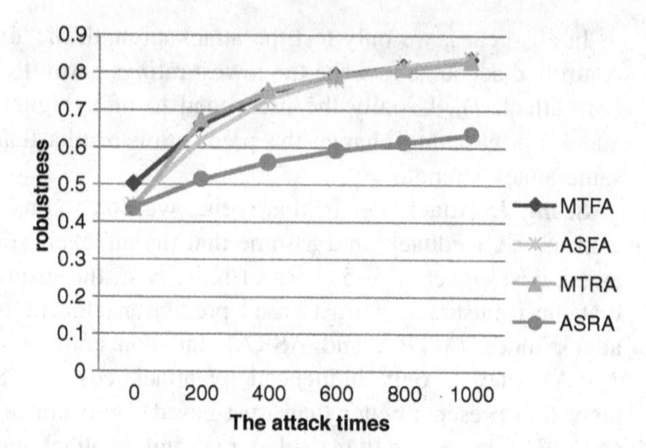

Fig. 7.3 Scene 2: the robustness change in direct attack

robustness. From the Fig. 7.4, we can see that the increase of false accounts has no significant influence to robustness. As compared, AS algorithm is influenced much more. In addition, compare to Fig. 7.2, Sybil attack has less influence than direct attack to algorithm robustness in the same cost c_G.

Scene 4: Attack T'_G and F_G. We set c_G as the basic cost, each account in U_T can create 100 false accounts. We test the tendency of robustness with users' feedback. It is similar to scene 3, we consider the attacker could easily create trust statement from U_T to U'_T, attack T_D begin from U'_T. The attack order and method is same as scene 2. From Fig. 7.5, we can see that trust-based prediction algorithm can recover quickly from attack through the feedback F_G from honest users.

Scene 5: Mixing attack. We set c_G as the basic cost, and set each account in U_T can create 50–250 distrust statement. Attack T_D begin from U'_T. This scene presents the attacker tried to shield high in-degree nodes in the trust network in order to reduce the weight of these nodes. Figure 7.6 shows that the number of distrust nodes has little influence to robustness.

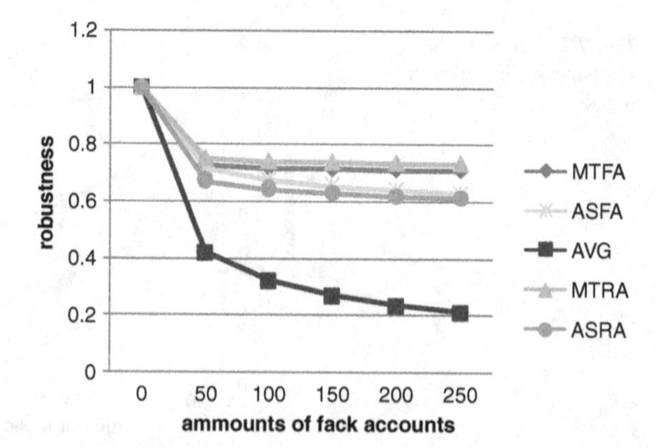

Fig. 7.4 Scene 3: the robustness change in Sybil attack

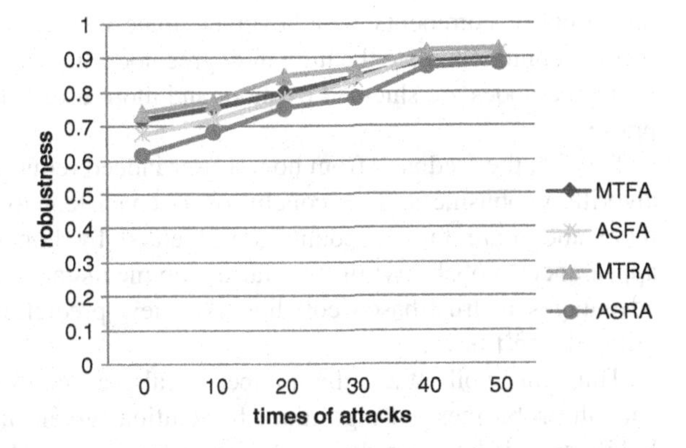

Fig. 7.5 Scene 4: the robustness change in Sybil attack

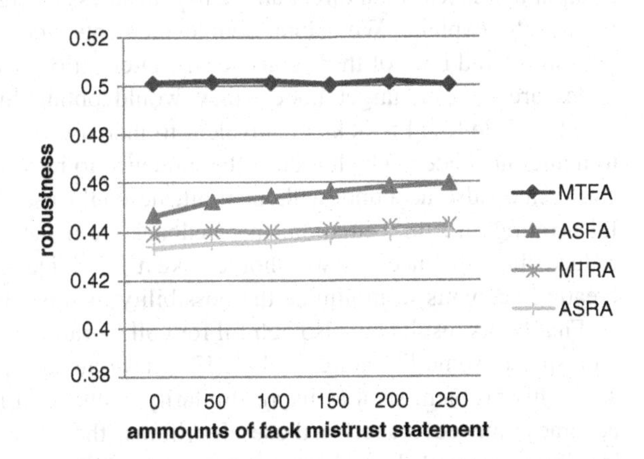

Fig. 7.6 Scene 5: the robustness change in mixing attack

7.7 Results Analysis

Through above experiments, we found that some seductive characteristics in the trust-based collective view prediction. In this section, we will analysis the reasons and give some good strategies for the weakness of algorithm.

First of all, the key factor to influence the robustness is the number of trust statement obtained by occupied accounts. We consider the reason is the characteristic of trust network. As a form of social network, some nodes in trust network have very high in-degree (see Chap. 5). These nodes are core hub in the network. They play as bridges for connecting other nodes. In addition, because of the short diameter of trust network, when we adopt limit depth breadth-first-search to search trust metric, these nodes would obtain high weights in results as they have high probability appear in neighbor sets. A few high in-degree nodes have the same effect as low in-degree nodes in robustness. This means if the attacker controls some high in-degree nodes, he could change the results significantly. Fortunately, when honest users discover false comments and provide feedback, the effect will be gradually reduced. In addition, we also notice that the users who write mass high quality comments will be given material rewards in Epinions.com. This method could stimulate the high in-degree nodes to keep good behavior, if the high in-degree nodes are shielded by more and more users, they will receive little good praise.

Second, the feedback from honest user indeed reduces the influence of attack to algorithm robustness. This conclusion is consistent to our common sense. With more and more false accounts are shielded by honest users, the weights and appearance probability of the false comments also reduce and it is another advantages in trust-based collaborative view prediction which collaborative filtering doesn't have.

Third, in Sybil attack, the number of false accounts cannot significantly influence the robustness of algorithm. In addition, given the cost c_G, the effect from Sybil attack is less than direct attack to robustness of algorithm. This phenomenon is easy to explain. We adopt double elements and single element prediction algorithm, and both of them consider the attenuation when trust transferred. If the nodes are near to target nodes, they would obtain high weights in prediction results, while Sybil attack is equivalent to increase the distance from target nodes to malicious nodes, which reduce the influence to results. However, if the attacker can create false accounts without limit, he could use different accounts to write false ratings. In this situation, the feedback from the honest users is helpless to reduce the influence, so we should take the advice in 7.5: set some rules for creating accounts to minimize the possibility of this situation happening.

Finally, its results are also helpful for collaborative filtering. For the problem of data sparse, some literature [9, 176, 177] have already proposed some methods to solve this problem such as using similarity connection in network to improve the accuracy and coverage. In those algorithms, the nodes with high degree in the similarity network have large influence to prediction results. How to avoid this weakness used by attacker needs our further research.

Chapter 8
Conclusions

This book compares and contrasts three different approaches to deal with web information process. They are web sentiment analysis, collaborative filtering and trust-based collective view prediction. Sentiment analysis takes text information as its algorithm input, while collaborative filtering requires rating matrix as its algorithm input. In order to cope with the noise data in rating matrix, we present a novel approach which is based on trust network.

Our work explores some research questions such as what is the relationship between trust and the similarity of users' opinions and how to describe the potential attacks and the ability to prevent attack of the system. With the intention to answer these questions, we form a methodology by creating a formal assessment framework, and by investigating some big dataset and conducting simulation experiments in real data.

Trust-based collective view prediction takes advantage of user trust statement and comment feedback, judges the reliability of the source of information to produce personalized prediction results, and resists the influence of the noisy data. It is an extension and development of traditional collective view prediction technology. This book focuses on this topic, and conducts some innovative work in the basic theory, prediction algorithms and analysis methods. Our major work and the contributions are as follows:

We establish a theory foundation to discuss and elaborate trust concept and its relationship with collective view prediction models. This book presents a clear definition of the differences and relations between trust and the similarity of users. Since only in a particular environment can the user similarity be used as a trusted source, the trust-based collective view prediction is essentially different from collaborative filtering. Earlier collective view prediction works with the term "trust", actually belongs to the field of collaborative filtering. In addition to the conceptual level, we also analyze the trust and user similarity quantitatively. In order to take advantage of the trust relationship data to predict the reasonable results, some connection must exist between the trust network structure and user insights. This book reveals this through statistical means. We found that trust degree and the distance between nodes in the network have a significant impact on

T. Luo et al., *Trust-Based Collective View Prediction*,
DOI: 10.1007/978-1-4614-7202-5_8,

user opinions. Therefore, nearest neighbor principle and trust weighted principle are two basic principles to build trust-based collective view algorithm. We also compare the similarities and differences in the structure of the trust network built by the Beta distribution function network and built based on trust statement. The analysis results show that the actual relationship of trust between the Beta distribution function can more accurately reflect the user input, that is user feedback in the comments, so that a reasonable trust relationship is built. The results also show that, compared with the former, trust network built by Beta distribution function can include more nodes. This coverage is useful for enhancing confidence-building measure. In the experiment, we use the sampled data from Epinions.com and public comment network in China. Furthermore, by comparing the results of differenct data sets, we found the correlation between trust metrics and user insights, and to take better advantage of this, we should allow the user to understand the role of the trust statement, and guide them to manage their own trust list more carefully.

Designing several effective and efficient prediction algorithms is not trivial work. We improved two trust metrics in the book, non-normalized MoleTrust [15] and normalized Appleseed [16] respectively, we also designed corresponding collective view prediction algorithm for them. The results of a comparative assessment showed two different trust metrics strategies have little effect on the prediction results. Inspired by the above algorithm, we also proposed a second-order Markov random walk model, which is better in prediction accuracy and prediction coverage than traditional collaborative filtering algorithms. We also tried a Bayesian model to fit more than one prediction algorithm, to further improve prediction accuracy. The experimental results show that the outputs of trust-based prediction algorithm and the second-order Markov model algorithm linearly related in a high degree. From another viewpoint, there is a close link between trust and user similarity.

In evaluating a collective view predicting modles and related algorithms, we not only take into account the accuracy and coverage, but also consider algorithms' robustness. We present an assessment framework and give formal description of attack, attack strength, algorithm robustness and common attack strategy for trust-based collective view prediction. The framework can be applied to the analysis of any collective view prediction algorithm which takes the trust network as input. In this framework, we evaluated two trust-based collective view prediction algorithms proposed in this study. The evaluation results show: First, the key factor that affect algorithm robustness is how much in-degree the attacker can get from the honest users in the trust network. Second, trust-based collective view prediction algorithm can effectively prevent the "behind-the-scenes attack". Third, trust-based collective view prediction algorithms can take advantage of feedback from honest users, and self-healing from the attack. We also propose two measures to enhance the robustness for the application of the algorithm in a real system, including adding incentives to the trust network node with high-degree and setting the newly created account minimum experience value.

The future direction of our work is to build a model for depicting the prediction algorithm behavior and the dynamic of trust network evolution, in oder to better detect potential attacks. The model-based prediction algorithm would play an important role in enhancing the efficiency of the algorithm for online prediction. These promising advantages also have compensations. Comparing to the prediction model based on the two-dimensional scoring matrix, adding an extra data dimension would be more difficult to build a practical model with low computational complexity. Since trust network is a special form of social networks, on the one hand, it has its own evolution patterns and properties; if we can discover their development laws and make use of them, we would be able to detect suspicious user behavior from trust network progress data. One the other hand, it has phenomena similar to social networks, such as the sudden change of network nodes clustering coefficient and the inconsistency of the growth rates between some subgraph and the global-graph. These issues still need to further discuss and explore.

Appendix
The Modified MoleTrust Algorithm

Input: Trust network G Source of measure u_x Trust degree threshold TT Max depth of measure maxdep

Output: Trust degrees and corresponding depth of all the reachable nodes of maxdep

Begin:

  set $i \leftarrow 1,\ REC \leftarrow \{\}$;

  $\forall u_y \in G_x :$ **set** $N_1 \leftarrow N_1 \cup \{u_y\},\ REC \leftarrow REC \cup \{(t_{xy}, depth)\}$;

  while $N_i \neq \{\}$ **and** $i < maxdep$ **do**

      set $TEMP \leftarrow \{\}$;

        for all $u_w \in N_i$ **do**

            if $t_{xw} > TT$ **then**

            for all $u_z \in G_w$ **do**

            if $(t_{xz}, depth) \notin REC$ **then**

              if $(t_{xz}, \omega_{xz}) \in TEMP$ **then**

                set $TEMP(z) \leftarrow (t_{xz} + t_{xw} \cdot t_{wz}, \omega_{xz} + t_{xw})$;   *1

            else

                set $TEMP(z) \leftarrow (t_{xw} \cdot t_{wz}, t_{xw})$;   *2

            end if

            end if

      end do

      end if

    end do

  $\forall (t_{xz}, \omega_{xz}) \in TEMP :$ **set** $REC(z) \leftarrow (t_{xz} / \omega_{xz}, i)$;

  $i \leftarrow i + 1$;

  end do

  return REC;

We modified MoleTrust algorithm that do not change trust network into a directed acyclic graph before trust measuring. In this way, the new algorithm can avoid losing some certain reachable path and improve the coverage of measuring.

T. Luo et al., *Trust-Based Collective View Prediction*,
DOI: 10.1007/978-1-4614-7202-5, © Springer Science+Business Media New York 2013

In the above pseudo code, N_i represents the neighborhood of depth *I* and it records the trust measurement results in the current depth. S_x represents the set of sub-sequent nodes of u_x in the trust network. *TEMP* represents the set of temporary results. It records the temporary measurement results as binary group (the total confidence, the current total weights), corresponding to the molecular and denominator of Eq. (4.7) respectively. In the Sect. 5, we do further improvement of MoleTrust in order to solve the problem of amplification of trust measurement. Eq. (5.2) can be achieved by replace the "*" line in the pseudo with the following code..

$$\textbf{set } num_{xz} \leftarrow num_{xz} + \frac{\omega_{xw}}{V^t_{\max}} \cdot t_{wz},$$

$$den_{xz} \leftarrow den_{xz} + 1; *1$$

$$\textbf{set } num_{xz} \leftarrow \frac{\omega_{xw}}{V^t_{\max}} \cdot t_{wz},\ den_{xz} \leftarrow 1; *2$$

The above code also needs to modify a little bit to use Beta function to build trust network. For more details, see Sect. A.1.

A.1 Modified MoleTrust Algorithm Used in Beta trust Network

Input: Trust network G Source of measure u_x Trust degree threshold TT Max depth of measure maxdep

Output: Trust degrees and corresponding depth of all the reachable nodes of maxdep

```
Begin:
    set i ← 1, N_1 ← {};
    ∀u_y ∈ S_x : set ω_xy ← t_xy , N_1 ← N_1 ∪ {ω_xy};
    while N_i ≠ {} and i < maxdep do
            set N_{i+1} ← {}, TEMP ← {};
            for all ω_xw ∈ N_i do
            if ω_xw > TT then
                  for all u_z ∈ S_w do
                     if ω_xz ∉ {N_1, N_2,..., N_i} then
                            if (p_xz, n_xz) ∈ TEMP then
                                 set TEMP(z) ← ω_xw ⊗ t_wz ⊕ TEMP(z);
                     else
                                 set TEMP(z) ← ω_xw ⊗ t_wz;
                           end if
                     end if
                     end do
           end if
```

The above algorithm uses the operator of faith theory to merge the trust degree. Hereω_{xz} represents both the truest degree which is expressed by binary group (p_{xz}, n_{xz}) and the corresponding real value of trust degree.

A.2 Modified Appleseed Algorithm

Input: Trust network G Source of measure u_x Trust degree threshold *TT* Max depth of measure *maxdep*

Output: Trust degrees and corresponding depth of all the reachable nodes of *maxdep*

Begin:
 set $in_0(x) \leftarrow 1$, $trust_0(x) \leftarrow 0$, $i \leftarrow 0$, $A_0 \leftarrow \{x\}$;
 repeat
  set $i \leftarrow i+1$, $A_i \leftarrow A_{i-1}$;
  $\forall w \in A_{i-1}$: **set** $in_i(w) \leftarrow 0$;
  for all $w \in A_{i-1}$ **do**
   set $trust_i(w) \leftarrow trust_{i-1}(w) + p \cdot in_{i-1}(w)$;
   if $in_{i-1}(w) > 0$ **then**
   for all $(w, z) \in G$ **do**
    if $z \notin A_i$ **then**
    set $A_i \leftarrow A_i \cup \{z\}$, $in_i(z) \leftarrow 0$, $trust_i(z) \leftarrow 0$;
    add link (z, x);
    set $\omega(z, x) \leftarrow 1$;
    end if
    set $\omega \leftarrow \omega(w, z) / \sum_{(w,z')\in G} |\omega(w, z')|$;
    set $in_i(z) \leftarrow in_i(z) + (1-p) \cdot \omega \cdot in_{i-1}(w)$;
     end do
    end if
   end do
 until ($i > maxdep$)
  return $\{trust_i(z) \mid z \in A_i\}$;

In order to prevent some nodes only receiving energy from other nodes but not sending energy, Appleseed [16] algorithm takes the following strategies. If there is not a trust statement between node u_z and measure source u_x, then add a virtual statement and give it high trust degree. Otherwise, create trust statement with the same rule and cover the original trust statement. The strategies we used are adding a virtual statement between $u_z$and measure source u_x directly and give it high trust degree. If there is already a trust statement between u_z and measure source u_x, then there are two trust statements between them in the new strategy, so that more energy can back into the nodes which are near to source node. We think it can reflect the "distance priority" strategy which is determined in Chap. 3 better. In order to facilitate the reader to contrast, the symbolic representation we used in the above pseudo code is similar with that in literature [16].

A_i represents the calculation energy sweep node set after i times iteration. (w, z) represents the arc from node $w$to node $z.\omega(w, z)$ is the weight of the arc. $in_i(x)$ band $trust_i(x)$ represent the injection energy of node x and energy already captured after i times integrations. The above algorithms assume that trust degree has been normalized to $[-1, 1]$ interval [1].

References

1. Liu, B., *Sentiment analysis and subjectivity.* A Chapter in Handbook of Natural Language Processing. 2010.
2. Breese, J., D. Heckerman, and C. Kadie, *Empirical analysis of predictive algorithms for collaborative filtering*, in *In Proceedings of the Fourteenth Conference on Uncertainty in Artificial Intelligence (UAI).* 1998: San Francisco, USA. p. 43–52.
3. Adomavicius, G. and A. Tuzhilin, *Toward the next generation of recommender systems: A survey of the state-of-the-art and possible extensions.* IEEE Transactions on Knowledge and Data Engineering, 2005. **17**(6): p. 734–749.
4. Das, A., M. Datar, and A. Garg, *Google news personalization: Scalable online collaborative filtering*, in *In Proceedings of WWW'07.* 2007: Banff, Alberta, Canada. p. 272-280.
5. Linden, G., B. Smith, and J. York, *Amazon.com recommendation: Item-to-item collaborative filtering.* IEEE Internet Computing, 2003: p. 76–80.
6. O'Mahony, M., N. Hurley, and G. Silvestre, *Promoting Recommendations: An Attack on Collaborative Filtering.* Database and Expert Systems Applications, 2002: p. 213-241.
7. O'Mahony, M., et al., *Collaborative recommendation: A robustness analysis.* ACM Transaction on Internet Technology, 2004. **4**(4): p. 344–377.
8. Melville, P., R.J. Mooney, and R. Nagarajan, *Content-Boosted Collaborative Filtering for Improved Recommendations*, in *Eighteenth national conference on Artificial intelligence.* 2002: Edmonton, Alberta, Canada p. 187-192.
9. Huang, Z., H. Chen, and D. Zeng, *Applying associative retrieval techniques to alleviate the sparsity problem in collaborative filtering.* ACM Transactions on Information Systems, 2004. **22**(1): p. 116-142.
10. O'Mahony, M., N. Hurley, and G. Silvestre, *An evaluation of neighborhood formation on the performance of collaborative filtering.* Artificial Intelligence Review, 2004. **21**(3-4): p. 215-228.
11. Burke, R., et al., *Classification features for attack detection in collaborative recommender systems*, in *Proceedings of the 12th ACM SIGKDD international conference on Knowledge discovery and data mining* 2006: New York, NY, USA p. 542-547
12. Mehta, B., T. Hofmann, and W. Nejdl, *Robust Collaborative Filtering.* Proceedings of the 2007 ACM conference on Recommender systems 2007: p. 49-56
13. Mehta, B. and W. Nejdl, *Attack-resistant Collaborative Filtering*, in *Proceedings of the 31st annual international ACM SIGIR conference on Research and development in information retrieval* 2008: New York, NY, USA. p. 75-82
14. Mehta, B. and T. Hofmann, *A Survey of Attack-Resistant Collaborative Filtering Algorithms.* Bulletin of the IEEE Computer Society Technical Committee on Data Engineering, 2008. **31**(3).
15. Massa, P. and P. Avesani, *Trust-aware recommender systems*, in *ACM conference on Recommender systems.* 2007: Minneapolis, Minnesota, USA.

T. Luo et al., *Trust-Based Collective View Prediction*,
DOI: 10.1007/978-1-4614-7202-5,

16. Ziegler, C.-N., *Towards decentralized recommender system.* 2005, Albert-Ludwigs: Freiburg
17. Golbeck, J., *Computing and applying trust in web-based social networks.* 2005, University of Maryland.
18. O'Donovan, J. and B. Smyth, *Trust in recommender systems.* Proceedings of the 10th international conference on Intelligent user interfaces, 2005: p. 167-174.
19. Papagelis, M., D. Plexousakis, and T. Kutsuras, *Alleviating the Sparsity Problem of Collaborative Filtering Using Trust Inferences.* In Proceedings of iTrust 2005, 2005: p. 224–239.
20. Weng, J., C. Miao, and A. Goh, *Improving Collaborative Filtering with Trust-based Metric.* WWW '05 Proceedings of the 14th international conference on World Wide Web 2006: p. 22 - 32
21. Mitchell, T.M., *Machine Learning* 1997: McGraw-Hill, Inc.
22. Blaze, M., J. Feigenbaum, and J. Lacy, *Decentralized Trust Management*, in *In Proceedings of the IEEE Symp. on Security and Privacy*. 1999: Washington, USA. p. 164.
23. Jøsang, A., C. Keserand, and T. Dimitrakos, *Can We Manage Trust?*, in *In Proceedings of the 3rd International Conference on Trust Management*. 2005: Paris.
24. Salton, G., *Automatic text processing*. 1988: Addison-Wesley Longman Publishing Co., Inc.
25. Balabanovic, M. and Y. Shoham, *Fab: Content-Based, Collaborative Recommendation. Comm.* Comm. ACM, 1997. **40**(3): p. 66-72.
26. Rocchio, J.J., *Relevance Feedback in Information Retrieval*, in *In The SMART Retrieval System: Experiments in Automatic Document Processing* 1971. p. 313-323
27. Littlestone, N. and M. Warmuth, *The Weighted Majority Algorithm.* Information and Computation, 1994. **108**(2): p. 212-261.
28. Mooney, R.J., P.N. Bennett, and L. Roy, *Book Recommending Using Text Categorization with Extracted Information.* In Proc. Recommender Systems Papers from 1998 Workshop, Technical Report WS-98-08, 1998.
29. Pazzani, M. and D. Billsus, *Learning and Revising User Profiles: The Identification of Interesting Web Sites.* Machine Learning, 1997. **27**: p. 313-331.
30. Hongyu, W., et al., *A Recommendation Algorithm based on Support Vector Regression.* Journal of The Graduate School of the Chinese Academy of Sciences, 2007. **24**(6).
31. Zhu, T., et al., *Goal-Directed Site-Independent Recommendations from Passive Observations.* In Proceedings of the Twentieth National Conference on Artificial Intelligence, 2005.
32. Resnick, P., et al., *Grouplens: An open architecture for collaborative filtering of netnews*, in *In Proceedings of ACM 1994 Conference on Computer Supported Cooperative Work.* 1994: Chapel Hill, NC.
33. Basu, C., H. Hirsh, and W. Cohen, *Recommendation as Classification: Using Social and Content-based Information in Recommendation.* Recommendation System, 1998.
34. Claypool, M., et al. *Combining Content-based and Collaborative Filtering in an Online Newspaper.* in *ACM SIGKDD International Conference on Knowledge Discovery and Data Mining*. 1999.
35. Pazzani, M., *A Framework for Collaborative, Content-based and Demographic Filtering.* Artificial Intelligence Review, 1999. **27**: p. 313-331.
36. A.I., S., J. Konstan, and J. Riedl, *E-Commerce Recommendation Applications.* Data Mining and Knowledge Discovery, 2001. **5**: p. 115-153.
37. Soboroff, I.M. and C. Nicholas, *Collaborative Filtering and the Generalized Vector Space Model.* In Proceedings of the 23rd Annual International Conference on Researech and Development in Information Retrieval (SIGIR), 2000.
38. L.H., U. and F. D.P., *Clustering Methods for Collaborative Filtering.* Recommendation System, 1998.
39. Billsus, D. and M. Pazzani, *User Modeling for Adaptive News Access.* User Modeling and User-Adapted Interaction, 2000. **10**(23): p. 147-180.

40. Tran.T and R. Cohen, *Hybird Recommender Systems for Electronic Commerce*. Knowledge-Based Electronic Markets, 2000.
41. N., G., et al., *Combining Collaborative Filtering with Personal Agents for Better Recommendations*. Artificial Intelligence 1999: p. 439-446.
42. Soboroff, I.M. and C. Nicholas. *Combining Content and Collaboration in Text Filtering*. in *International Conference on Artificial Intelligence Workshop: Machine Learning for Information Filtering*. 1999.
43. A., P., et al., *Probabilistic Models for Unified Collaborative and Content-Based Recommendation in Sparse-Data Environments*, in *Proc. 17th Conf. Uncertainty in Artificial Intelligence*. 2001.
44. A.I., S., et al., *Methods and Metrics for Cold-Start Recommendations*, in *Proc. 25th Ann. Int'l ACM SIGIR Conf.* 2002.
45. M., C., et al. *Bayesian Mixed-Effects Models for Recommender Systems*. in *Proc. ACM SIGIR '99 Workshop Recommender Systems: Algorithms and Evaluation*. 1999.
46. Liu, B., *Sentiment analysis*, in *A presentation on 5th Text Analytics Summit*. 2009: Boston.
47. Pang, B. and L. Lee, *Opinion mining and sentiment analysis*. Foundations and Trends in Information Retrieval, 2008. **2**(1-2): p. 1-135.
48. Jindal, N. and B. Liu, *Opinion spam and analysis*, in *In Proceedings of the international conference on Web search and web data mining*. 2007: Palo Alto, California, USA. p. 219-230.
49. Leskove, J., *Dynamics of large networks*. 2008, Carnegie Mellon University.
50. Abello, J., A.L. Buchsbaum, and J. Westbrook, *A functional approach to external graph algorithms*, in *In Proceedings of the 6th Annual European Symposium on Algorithms*. 1998: Springer-Verlag. p. 332-343.
51. Kleinberg, J.M., et al., *The web as a graph: Measurements, models and methods*, in *In COCOON '99: Proceedings of the International Conference on Combinatorics and Computing*. 1999: Tokyo, Japan
52. Broder, A., et al., *Graph structure in the web: experiments and models*. In WWW'00: Proceedings of the 9th international conference on World Wide Web, 2000.
53. Barabasi, A.-L. and R. Albert, *Emergence of scaling in random networks*. Science, 1999. **286**: p. 509-512.
54. Huberman, B.A. and L.A. Adamic, *Growth dynamics of the world-wide web*. Nature, 1999. **399**: p. 131.
55. Kumar, S.R., et al., *Trawling the web for emerging cybercommunities*. Computer Networks, 1999. **31**(11-16): p. 1481-1493.
56. Faloutsos, M., P. Faloutsos, and C. Faloutsos, *On power-law relationships of the internet topology*. Proceedings of the conference on Applications, technologies, architectures, and protocols for computer communication, 1999: p. 251-262.
57. Redner, S., *How popular is your paper? An empirical study of the citation distribution*. European Physical Journal B, 1998. **4**: p. 131-134.
58. Chakrabarti, D., Y. Zhan, and C. Faloutsos, *R-mat: A recursive model for graph mining*. In SDM '04: SIAM Conference on Data Mining, 2004.
59. Albert, R. and A.-L. Barabasi, *Statistical mechanics of complex networks*. Reviews of Modern Physics, 2002. **74**(1): p. 47-97.
60. Pennock, D.M., et al., *Winners don't take all: Characterizing the competition for links on the Web*. Proceedings of the National Academy of Sciences, 2002. **99**(8): p. 5207-5211.
61. Bi, Z., C. Faloutsos, and F. Korn, *The DGX distribution for mining massive, skewed data*. In KDD'01: Proceedings of the 6th ACM SIGKDD international conference on Knowledge discovery and data mining, 2001: p. 17-26.
62. Newman, M.E.J., *Power laws, pareto distributions and zipf's law*. Contemporary Physics, 2005. **46**: p. 323-351.
63. Tauro, S.L., et al., *A simple conceptual model for the internet topology*. In GLOBECOM'01: Global Telecommunications Conference, 2001. **3**: p. 1667-1671.

64. Milgram, S., *The small-world problem.* Psychology Today, 1967. **2**: p. 60-67.
65. Albert, R., H. Jeong, and A.-L. Barabasi, *Diameter of the world-wide web.* Nature, 1999. **401**: p. 130-131.
66. Bollobas, B. and O. Riordan, *The diameter of a scale-free random graph.* Combinatorica, 2004. **24**(1): p. 5-34.
67. Chung, F. and L. Lu, *The average distances in random graphs with given expected degrees.* The average distances in random graphs with given expected degrees, 2002. **99**(25): p. 15879–15882.
68. Watts, D.J. and S.H. Strogatz, *Collective dynamics of 'small-world' networks.* Nature, 1998. **393**: p. 440-442.
69. Dorogovtsev, S.N., A.V. Goltsev, and J.F.F. Mendes, *Pseudofractal scale-free web.* Physical Review E, 2002. **65**(6): p. 66-122.
70. Clauset, A., C. Moore, and M.E.J. Newman, *Hierarchical structure and the prediction of missing links in networks.* Nature, 2008. **453**(7191): p. 98-101.
71. Newman, M.E.J., *Detecting community structure in networks.* The European Physical Journal B, 2004. **38**: p. 321-330.
72. Danon, L., et al., *Comparing community structure identification.* Journal of Statistical Mechanics: Theory and Experiment, 2005. **29**(9).
73. Palla, G., et al., *Uncovering the overlapping community structure of complex networks in nature and society.* Nature, 2005. **435**(7043): p. 814-818.
74. Milo, R., et al., *Network Motifs: Simple Building Blocks of Complex Networks.* Science, 2002. **298**(5594): p. 824-827.
75. Alon, U., *Network motifs: theory and experimental approaches.* Nature Reviews Genetics, 2007. **8**(6): p. 450-461.
76. Palmer, C.R., P.B. Gibbons, and C. Faloutsos, *Anf: a fast and scalable tool for data mining in massive graphs.* In KDD'02: Proceedings of the 8th ACM SIGKDD international conference on Knowledge discovery and data mining, 2002: p. 81-90.
77. Leskovec, J., J. Kleinberg, and C. Faloutsos, *Graphs over Time: Densification Laws, Shrinking Diameters and Possible Explanations.* ACM SIGKDD International Conference on Knowledge Discovery and Data Mining, 2005.
78. J, G., *The Dynamics of Web-based Social Networks: Membership, Relationships, and Change.* First Monday, 2007. **12**(11).
79. Cooper, C. and A. Frieze, *A general model of web graphs.* Random Structures and Algorithms, 2003. **22**(3): p. 311-335.
80. Kumar, R., et al., *Stochastic models for the web graph.* In FOCS'00: Proceedings of the 41st Annual Symposium on Foundations of Computer Science, 2000.
81. Wasserman, S. and P. Pattison, *Logit models and logistic regressions for social networks.* Psychometrika, 1996. **60**: p. 401-425.
82. Wiuf, C., et al., *A likelihood approach to analysis of network data.* Proceedings of the National Academy of Sciences, 2006. **103**(20): p. 7566-7570.
83. Han, J. and M. Kamber, *Data Mining: Concepts and Techniques*. 2005: Morgan Kaufmann.
84. Easley and David, *Networks, Crowds and Markets: Reasoning about a Highly Connected Workld.* 2 ed. 2010: Cambridge University.
85. Rogers, E.M., *Diffusion of Innovations*. 4 ed. 1995: Free Press.
86. Domingos, P. and M. Richardson, *Mining the network value of customers*, in *In KDD'01: Proceedings of the 7th ACM SIGKDD international conference on Knowledge discovery and data mining*. 2001.
87. Gomi, H., et al., *A delegation framework for federated identity management*, in *In Proceedings of the Workshop on Digital Identity Management*. 2005: Virginia, USA. p. 94-103.
88. Kumar, R., et al., *On the bursty evolution of blogspace.* In WWW'02: Proceedings of the 11th international conference on World Wide Web, 2003: p. 568-576.

89. Adamic, L.A. and N. Glance, *The political blogosphere and the 2004 u.s. election: divided they blog.* In LinkKDD'05: Proceedings of the 3rd international workshop on Link discovery, 2005: p. 36-43.
90. Adar, E. and L.A. Adamic, *Tracking information epidemics in blogspace.* In Web Intelligence, 2005: p. 207-214.
91. Gruhl, D., et al., *Information diffusion through blogspace.* In WWW'04: Proceedings of the 13th international conference on World Wide Web, 2004: p. 491-501.
92. Abdul-Rahman, A., *The PGP Trust Model.* EDI-Forum, 1997.
93. X.509, I.-T.R., *The Directory - Public-Keyand Attribute Certificate Frameworks.* 2000.
94. Winsborough, W.H., K.E. Seamons, and V.E. Jones, *Automated trust negotiation*, in *In: DARPA Information Survivability Conf. and Exposition.* 2000: New York, USA. p. 88-102.
95. Bhatti, R., E. Bertino, and A. Ghafoor, *An Integrated Approach to Federated Identity and Privilege Management in Open Systems.* Communications of the ACM, 2007. **50**(2): p. 81-87.
96. Sinnott, R., et al. *Supporting Decentralized, Security Focused Dynamic VirtualOrganizations across the Grid.* in *In Proceedings of Second International Conference on e-Science and Grid Computing.* 2006.
97. Chen, S., et al., *A Framework for Managing Access of Large-Scale Distributed Resources in a Collaborative Platform.* CODATA Data Science Journal, 2008. **7**: p. 137-147.
98. Grandison, T.W.A., *Trust Management for Internet Applications.* 2003, University of London.
99. Meulpolder, M., et al., *BarterCast: A practical approach to prevent lazy freeriding in P2P networks*, in *In Proceedings of 2009 IEEE International Symposium on Parallel & Distributed Processing.* 2009: Rome, Italy, . p. 1-8.
100. Adar, E. and B.A. Huberman, *Free Riding on Gnutella.* First Monday, 2000. **5**.
101. Schlosser, A., M. Voss, and L. Bruckner, *Comparing and Evaluating Metrics for Reputation Systems by Simulation.* In Proc. IAT Workshop on Reputation in Agent Societies, 2004.
102. Josang, A., R. Ismail, and C. Boyd, *A Survey of Trust and Reputation Systems for Online Service Provision.* Decision Support Systems, 2007. **43**(2): p. 618-644.
103. Casella, G. and R.L. Berger, *Statistical Inference.* 1990: Duxbury Press.
104. Josang, A., *A logic for uncertain probabilities.* International Journal of Uncertainty, Fuzziness and Knowledge-Based Systems, 2001. **3**: p. 279-311.
105. Konstan, J.A., et al., *GroupLens: applying collaborative filtering to Usenet news.* Communications of the ACM 1997. **40**(3): p. 77 - 87
106. Hill, W., et al., *Recommending and evaluating choices in a virtual community of use.* CHI '95 Proceedings of the SIGCHI Conference on Human Factors in Computing Systems 1995: p. 4-201.
107. Shani, G., D. Heckerman, and R. Brafman, *An mdp-based recommender system.* Journal of Machine Learning Research, 2005. **6**: p. 1265–1295.
108. Goldberg, K., et al., *Eigentaste: A Constant Time Collaborative Filtering Algorithm.* Information Retrieval J, 2001. **4**(2): p. 133-151.
109. Terveen, L., et al., *PHOAKS: A System for Sharing Recommendations.* Comm.ACM SIGKDD International Conference on Knowledge Discovery and Data Mining, 1997. **40**(3): p. 59-62.
110. Sarwar, B., et al., *Item-based collaborative filtering recommendation algorithms.* In Proceedings of WWW'01, Hong Kong, 2001: p. 285–295.
111. Delgado, J. and N. Ishii, *Memory-Based Weighted-Majority Prediction for Recommender Systems.* Proc. ACM SIGIR '99 Workshop Recommender Systems: Algorithms and Evaluation, 1999.
112. Rennie, J.D.M. and N. Srebro, *Fast maximum margin matrix factorization for collaborative prediction.* ICML '05: Proceedings of the 22nd international conference on Machine learning, 2005: p. 713–719.

113. Koren, Y., *Factorization meets the neighborhood: a multifaceted collaborative filtering model.* KDD '08: Proceeding of the 14th ACM SIGKDD international conference on Knowledge discovery and data mining, 2008: p. 426–434.
114. Tak′acs, G., et al., *Major components of the gravity recommendation system.* SIGKDD Explor. Newsl., 2007. **9**(2): p. 80–83.
115. Fouss, F., et al., *Random-Walk Computation of Similarities between Nodes of a graph with application to collaborative filtering.* IEEE Transactions on Knowledge and Data Engineering, 2007. **19**(3): p. 355-369.
116. Haveliwala, T.H., *Topic-sensitive PageRank: a context-sensitive ranking algorithm for Web search* IEEE Transactions on Knowledge and Data Engineering, 2003. **15**(4): p. 784 - 796.
117. *Global advertising consumers trust real friends and virtual strangers the most.*2009; Available from: http: //blog.nielsen.com/nielsenwire/consumer/global-advertising-consumers-trust-real-friends-and-virtual-strangers-the-most/.
118. Zhou, J. and T. Luo, *Modeling Learners and Contents in Academic-oriented Recommendation Framework*, in *Document Analysis Systems - DAS*. 2011.
119. Zhou, J. and T. Luo, *An Academic Resource Collaborative Recommendation Algorithm Based on Content Correlation.* Journal of The Graduate School of the Chinese Academy of Sciences, 2013.
120. Ding, Y. and X. Li, *Time weight collaborative filtering*, in *International Conference on Information and Knowledge Management - CIKM*. 2005. p. 485-492.
121. Zimdars, A., D.M. Chickering, and C. Meek, *Using temporal data for making recommendations*, in *Uncertainty in Artificial Intelligence-UAI*. 2001. p. 580-588.
122. Lathia, N., S. Hailes, and L. Capra, *Temporal collaborative filtering with adaptive neighbourhoods*, in *Research and Development in Information Retrieval - SIGIR*. 2009. p. 796-797.
123. Koren, Y., *Collaborative filtering with temporal dynamics.* Communications of The ACM - CACM, 2009. **53**(4): p. 447-456.
124. Zhou, J. and T. Luo, *Academic Recommendation on Graph with Dynamic Transfer Chain*, in *Social Computing and its Applications*. 2012.
125. Zhang, Y., D.Shen, and C.Baudi, *Sentiment Analysis in Practice.* ICDM2011 tutorial, 2011.
126. *SentiWordNet*. Available from: http: //patty.isti.cnr.it/ ~ esuli/software/SentiWordNet.
127. Fellbaum, C., *Wordnet: An Electronic Lexical Database*. 1998: MIT Press.
128. Yu, H. and V. Hatzivassiloglou, *Towards Answering Opinion Questions: Separating Facts from Opinions and Identifying the Polarity of Opinion Sentences*, in *Proceedings of the Conference on Empirical Methods in Natural Language Processing (EMNLP)*. 2003.
129. Lu, Y., et al. *Automatic Construction of a Context-Aware Sentiment Lexicon: An Optimization Approach.* in *Proceedings of the 20th international conference on World Wide Web (WWW)*. 2011.
130. Pang, B., L. Lee, and S. Vaithyanathan, *Thumbs up? Sentiment Classification Using Machine Learning Techniques*, in *Proceedings of the Conference on Empirical Methods in Natural Language Processing (EMNLP)*. 2002.
131. Pang, B. and L. Lee, *Seeing Stars: Exploiting Class Relationships for Sentiment Categorization with Respect to Rating Scales*, in *Proceedings of the Association for Computational Linguistics (ACL)*. 2005.
132. Turney, P., *Thumbs Up or Thumbs Down? Semantic Orientation Applied to Unsupervised Classification of Reviews.* Proceedings of the Association for Computational Linguistics (ACL), 2002.
133. Blitzer, J. and M. Dredze, *Biographies, Bollywood, Boom-boxes and Blenders: Domain Adaptation for Sentiment Classification*, in *Proceedings of the Association for Computational Linguistics (ACL)*. 2007.
134. Goldberg, D., et al., *Using collaborative filtering to weave an information tapestry.* Communications of the ACM, 1992. **35**: p. 61-70.

135. Hofmann, T., *Latent semantic models for collaborative filtering.* ACM Transactions on Information Systems, 2004. **22**(1): p. 89–115.
136. Golder, S. and B. Huberman, *Usage patterns of collaborative tagging systems.* Journal of Information Science, 2006. **32**: p. 198-208.
137. Zollers, A., *Emerging Motivations for Tagging: Expression, Performance, and Activism*, in *WWW 2007*. 2007.
138. Durao, F. and Dolog, *A personalized tag-based recommendation in social web systems*, in *Workshop on Adaptation and Personalization for Web 2.0, UMAP'09*. 2009.
139. Gedikli, F. and D. Jannach, *Rating items by rating tags.* Systems and the Social Web at ACM 2010.
140. Zhao, S., et al., *Improved recommendation based on collaborative tagging behaviors*, in *Intelligent User Interfaces - IUI*. 2008. p. 413-416.
141. Shepitsen, A., et al., *Personalized recommendation in social tagging systems using hierarchical clustering*, in *Conference on Recommender Systems - RecSys*. 2008. p. 259-266.
142. Xu, G., et al., *SemRec: A Semantic Enhancement Framework for Tag based Recommendation*, in *AAAI Conference on Artificial Intelligence*. 2011.
143. Leung, C., et al., *A probabilistic rating inference framework for mining user preferences from reviews.* World Wide Web - WWW, 2011. **14**(2): p. 187-215.
144. Wiebe, J., et al., *Learning subjective language.* Computational Linguistics - COLI, 2004. **30**(3): p. 277-308.
145. Whitelaw, C., et al., *Red Opal: product-feature scoring from reviews*, in *ACM Conference on Electronic Commerce - EC*. 2007. p. 182-191.
146. Kamps, J. and M. Marx, *Words with attitude*, in *In Proc. of the First International Conference on Global WordNet*. 2002. p. 332–341.
147. Zhuang, L., F. Jing, and X. Zhu, *Movie review mining and summarization*, in *In Proceedings of the 15th ACM international conference o n Information and knowledge management*. 2006. p. 43-50.
148. Heymann, P., D. Ramage, and H. Molina, *Social tag prediction*, in *Research and Development in Information Retrieval - SIGIR*. 2008. p. 531-538.
149. Sen, S., et al. *The quest for quality tags.* in *International Conference on Supporting Group Work - GROUP*. 2007.
150. Sztompka, P., *Trust: A Sociological Theory*. 2004: Cambridge University.
151. Gambetta, D., *Can We Trust Trust?*, in *Trust: Making and Breaking Cooperative Relations*, D. Gambetta, Editor. 1990, Basil Blackwell. Oxford. p. 213–238.
152. McKnight, D.H. and N.L. Chervany, *The Meanings of Trust*, in *Technical Report MISRC Working Paper Series 96-04*. 1996, Management Information Systems Research Center.
153. Massa, P. and P. Avesani, *Trust metrics on controversial users: balancing between tyranny of the majority and echo chambers.* Chapter in Special Issue on Semantics of People and Culture, International Journal on Semantic Web and Information Systems 2007.
154. Guha, R., *Open rating systems*, in *Stanford Knowledge Systems Laboratory*. 2003: CA, USA, Tech. Rep.
155. Lin, C., et al., *Enhancing Grid Security with Trust Management.* In Proceedings of the IEEE International Conference on Services Computing (SCC'04), 2004.
156. Abdul-Rahman, A. and S. Hailes, *Supporting trust in virtual communities*, in *In Proceedings of the 33rd Hawaii International Conference on System Sciences*. 2000: Maui, HI, USA.
157. Jensen, C., J. Davis, and S. Farnham, *Finding others online: Reputation systems for social online spaces.*, in *In Proceedings of the SIGCHI Conference on Human Factors in Computing Systems. ACM Press*. 2002: Minneapolis, MN, USA. p. 447-454.
158. Josang, A., *The beta reputation system.* In Proceedings of the 15th Bled Electronic Commerce Conference, 2002.
159. Jin, R., et al., *Collaborative filtering with decoupled models for preferences and ratings*, in *In Proceedings of CIKM'03*. 2003: New Orleans, Louisiana, USA.

160. Beyer, K., et al., *When is nearest neighbor meaningful*, in *In Proc. 7th Int. Conf. Database Theory*. 1999. p. 217–235.
161. Wang, J., A. Vries, and M. Reinders, *Unifying user-based and item-based collaborative filtering approaches by similarity fusion*, in *In Proceedings of SIGIR'06*. 2006: Seattle, Washington, USA.
162. Pennock, D.M., et al., *Collaborative filtering by personality diagnosis: A hybrid memory- and model-based approach*, in *In Proceedings of the Sixteenth Conference on Uncertainty in Artificial Intelligence (UAI)*. 2000: Stanford, CA. p. 473–480.
163. Levien, R., *Attack-resistant trust metrics*. 2004, University of California at Berkeley: Berkeley, CA, USA.
164. Page, L., et al., *The pagerank citation ranking: Bring order to the web*. 1999, Stanford InfoLab, Stanford, CA, USA, Tech. Rep.
165. Quillian, R., *Semantic memory*. In Semantic Information Processing, M. Minsky, Ed. MIT Press, Boston, MA, USA, 1968: p. 227-270.
166. Kamvar, S., M. Schlosser, and H. Garcia-Molina, *The eigentrust algorithm for reputation management in p2p networks*, in *In Proceedings of WWW'03*. 2003: Budapest, Hungary. p. 640–651.
167. Jeh, G. and J. Widom, *Simrank: A measure of structuralcontext similarity*, in *In Proceedings of KDD'02, Edmonton*. 2002: Alberta, Canada. p. 538-543.
168. Sun, J., et al., *Neighborhood formation and anomaly detection in bipartite graphs*. In Proceedings of ICDM'05, 2005.
169. Clements, M., A.d. Vries, and M. Reinders, *Optimizing single term queries using a personalized markov random walk over the social graph*. In Proceedings of ESAIR'08, 2008.
170. Collins, A.M. and E.F. Loftus, *A spreading activation theory of semantic processing*. Psychology Today, 1975. **82**(6): p. 407-428.
171. Smith, E., et al., *Atkinson and Hilgards's Introduction to Psychology*. Thomson Learning, Boston, MA, USA, 2003.
172. Bollen, J., H. Vandesompel, and L.M. Rocha, *Mining associative relations from website logs and their application to context-dependent retrieval using spreading activation*. In Proceedings of the Workshop on Organizing Web Space (WOWS), 1999.
173. Crestani, F. and P.L. Lee, *Searching the web by constrained spreading activation*. Inf. Proc. Manage, 2000. **36**: p. 585-605.
174. Pirolli, P., J. Pitkow, and R. Rao, *Silk from a sow's ear: Extracting usable structures from the web*. In Proceedings of the ACMCHI 96 Conference on Human Factors in Computing Systems, 1996: p. 118-125.
175. Ceglowski, M., A. Coburn, and J. Cuadrado, *Semantic search of unstructured data using contextual network graphs*. 2003, National Institute for Technology and Liberal
176. Aggarwal, C.C., et al., *Horting hatches an egg: A new graphtheoretic approach to collaborative filtering*, in *In Proceedings of the 5th ACM SIGKDD Conference on Knowledge Discovery and Data Mining (KDD'99)*. 1999: San Diego, Calif. p. 201-212.
177. Mirza, B.J., *Jumping connections: A graph-theoretic model for recommender systems*, in *Computer Science Department*. 2001, Virginia Polytechnic Institute and state university.
178. Xiang, L. and QingYang, *Time-Dependent Models in Collaborative Filtering Based Recommender System*, in *Web Intelligence and Intelligent Agent Technologies*. 2009. p. 450-457.
179. Chen, L.-H., *Enhancement of Student Learning Performance using Personalized Diagnosis and Remedial Learning System*. Computers & Education, 2011. **56**(1): p. 289-299.
180. Klašnja-Milićević, A., et al., *E-Learning personalization based on Hybrid Recommendation strategy and Learning Style Identification*. Computers & Education, 2011. **56**(3): p. 885-899.

181. Tang, T.Y. and G. McCalla, *Smart Recommendation for an Evolving E-Learning System*, in *Proceedings of the IJCAI Workshop on Machine Learning in Information Filtering*. 2003. p. 86-91.
182. McCalla, G., *The ecological approach to the design of e-learning environments: purpose-based capture and use of information about learners.* Journal of Interactive media in Education, 2004. **7**
183. Drachsler, H., H.G.K. Hummel, and R. Koper, *Personal Recommendation Systems for Learners in Lifelong Learning Networks: the Requirements.* Techniques and Model. International Journal of Learning Technology, 2008. **3**: p. 404-423
184. Lee, J., K. Lee, and J. G.Kim. *Personalized academic research paper recommendation system.* Available from: http: //www.cc.gatech.edu/~jkim693/projects/recommendation.pdf.
185. Gipp, B., J. Beel, and C. Hentschel, *Scienstein: A Research Paper Recommender System*, in *In Proceedings of the International Conference on Emerging Trends in Computing (ICETiC'09)*. 2009. p. 309–315.
186. P. Chebotarev and E. Shamis, *The Matrix-Forest Theorem andMeasuring Relations in Small Social Groups*, Automation and Remote Control, vol. 58, no. 9, pp. 1505-1514, 1997.
187. P. Chebotarev and E. Shamis, *On Proximity Measures for GraphVertices*, Automation and Remote Control, vol. 59, no. 10, pp. 1443-1459, 1998.
188. S. Baluja, R. Seth, D. Sivakumar, Y. Jing, J. Yagnik, S. Kumar, D. Ravichandran, and M. Aly. *Video suggestion and discovery for youtube: taking random walks through the view graph.* In WWW '08, pp. 895–904, 2008.
189. Z. Huang, W. Chung, and H. Chen. *A graph model for e-commerce recommender systems.* J. Am. Soc. Inf. Sci. Technol., vol.55, pp. 259–274, 2004.
190. M. Gori, A. Pucci. *Research paper recommender systems: A random-walk based approach.* In VI'06, pp.778-781, 2006
191. A. Zimdars, D. M. Chickering, and C. Meek. *Using temporal data for making recommendations.* In UAI '01, pp. 580–588, 2001.
192. N. Lathia, S. Hailes, and L. Capra. *Temporal collaborative filtering with adaptive neighbourhoods.* In SIGIR '09, pp. 796–797, 2009.
193. http: //www.citeulike.org/home
194. http: //www.citeulike.org/faq/data.adp
195. G. Karypis. *Evaluation of item-based top-n recommendation algorithms.* In CIKM '01, pp. 247–254, 2001
196. J. Sun, D. Tao, and C. Faloutsos. *Beyond streams and graphs: dynamic tensor analysis.* In KDD '06, pp. 374–383, 2006.

Index

T. Luo et al., *Trust-Based Collective View Prediction*,
DOI: 10.1007/978-1-4614-7202-5,

Printed in the United States
By Bookmasters